Our Dynamic Earth: A Primer

Matthew R. Bennett

Our Dynamic Earth: A Primer

 Springer

Matthew R. Bennett
Faculty of Science and Technology
Bournemouth University
Poole, UK

ISBN 978-3-030-90353-4   ISBN 978-3-030-90351-0   (eBook)
https://doi.org/10.1007/978-3-030-90351-0

This Springer imprint is published by the registered company Springer Nature Switzerland AG
The registered company address is: Gewerbestrasse 11, 6330 Cham, Switzerland

To
Andrew F. Bennett
Father, Geographer, Rambler, Photographer,
and Politician

Preface

A primer is a readable introduction to a subject, more technical than a piece of popular science, but less detailed than a textbook. It aims to give the reader a platform in a subject with which they may be unfamiliar, so that they can proceed simultaneously, or sequentially, to more advanced texts and information. Ideally the primer should have something for those without any knowledge, while also challenge and entertain those who do. Not quite bedtime reading, but a step in that direction.

I have for many years taught an undergraduate physical geography unit at Bournemouth catering for students with different backgrounds. Some have never done geography since their early teens, while others have done it more recently. This range of backgrounds can be challenging, and this primer was conceived to help with this diversity of knowledge. Unlike a textbook, it is meant to be read sequentially from the first to last page and is not a substitute for more traditional textbooks, but a compliment to them.

The primer was first written in the summer of 2020 as a positive outcome of the Covid-19 pandemic with the aim of helping students through the trials of remote teaching during lockdown. It is based on the syllabus that I teach, which is no doubt similar to many introductory physical geography units, but there will be some differences, and it is hard to be inclusive of all subjects in a manageable text. If I have missed something out you consider vital, then let me know and I will correct it in a future edition.

Along the way, there are various tasks and suggestions for additional reading which can be ignored or used as prompts to make extra notes or delve into the research literature. I hope that you find this primer of use.

Poole, UK   Matthew R. Bennett
Summer 2021

Prologue: Why Physical Geography?

Imagine that you are a first-year student sitting fresh faced in your first physical geography class. Some of you will be excited having done physical geography before and enjoyed it, others will be saying, 'I only like human geography, the physical stuff is boring!' Perhaps others will be saying, 'I am an ecologist, why do I have to learn this geology stuff?' All are valid viewpoints.

I have spent my life as a geographer cum geologist, working in the high arctic on glacial processes, reconstructing our Ice Age past, studying the geography of human evolution in Africa, and applying geomorphological expertise to the study of forensic footprints at crime scenes. If you make a brief survey of the news bulletins, it will quickly become apparent that we must overcome huge environmental challenges as the Earth's climate continues to change. It is hardly surprising therefore that I believe that world's leaders and decision-makers all need to be both scientifically and geographically literate. Geography matters, and understanding how basic Earth systems operate is essential to us all. Let me try and show you why.

Roll the clock forward and imagine that you are now working for an aid organisation coordinating humanitarian relief. The news breaks of a major earthquake in northern Pakistan. You have to mobilise people and resources and get them to the epicentre fast. The questions flood in: what is the terrain like, what is the vegetation like, what is the climate and weather like, and where are the transport lines most vulnerable to aftershocks? This is just a small selection of questions – Google Earth and the Internet have their limits. Later you may be asked to advise on rebuilding lost infrastructure or improving disaster and emergency planning. All these questions are underpinned by physical geography.

If you do not like this scenario, image yourself as a conservation worker in Africa saving the rhino. The rhino is a product of its environment, the distribution of soil and food resources and its movements is limited by the local terrain. Climate change and local weather patterns all play a part in its survival even before we consider the social and cultural aspects that lead to its predation by poachers.

I could go on. Understanding the Earth's surface terrain, its shape, composition, and the processes that formed it in the past and that shape it now and will in the future is fundamental to almost all human interaction with the planet we live on.

Table 1 Some of the main sub-disciplines in physical geography

Geographical sub-disciplines
Geomorphology – shape of the Earth's surface and processes by which it is shaped, both at the present as well as in the past. It is closely linked to geology.
Hydrology – the distribution, movement, and quality of water on the land surface and in the soils and rocks near the surface. Ground water hydrology is known as geo-hydrology.
Glaciology – study of the Earth's current glaciers and ice sheets (cryosphere). It is closely associated with Quaternary Science.
Biogeography – study of the geographic patterns of species distribution and the processes that result in these patterns.
Climatology – study of the Earth's climate or weather patterns that predominate at a location, distinct from *meteorology* which is the study of day-to-day weather.
Pedology or Soil Science – the study of soils in their natural environment.
Oceanography – the study of the Earth's oceans and seas; many people would recognise this as a discipline in its own right.
Quaternary Science – is the inter-disciplinary study of the Quaternary period, which encompasses the last 2.8 million years. This includes understanding past climates, landscape changes, ice sheets, and the mechanisms of both climate and environmental change.
Geomatics – is the collection and processing of geographically relevant 'big-data' from satellites and Earth observation systems.
Environmental Geography – this focuses on the interaction of humans and the natural world. In some respects, it lies at the interface between human and physical geography.

That is what physical geography is about. It is the foundation of environmental and ecological science, a key component of geology, and therefore our understanding of Earth's history and our past. That is why all those interested in ecology, geography, and the environment need to be versed in fundamental Earth systems.

Physical geography is the study of the processes that shape the Earth's surface, the animals and plants that inhabit it, and their spatial distribution (Table 1). This surface lies at the interface between the lithosphere and the atmosphere and is shaped by both. Its study is by definition multi-disciplinary, drawing on geology and meteorology, and is fundamental to understanding the ecology and biogeography our planet.

As a discipline, it emerged in the mid- to late-1800s with geomorphologists dominating the discipline at first (Table 1). The emphasis was on the description of landscapes, climates, and biomes. Ideas of environmental determinism dominated in which landscapes in particular were seen as part of development cycles.

Physical geography along with human geography underwent a radical period of quantification in the late 1950s and early 1960s, known as the Quantitative Revolution. In geomorphology, there was a shift from the description of landforms to process-based experimentation on the mechanism by which landforms were formed. What followed was a massive growth in research and intense disciplinary specialisation around five broad themes: geomorphology, climatology, biogeography, soil science, and Quaternary environmental change.

Today, physical geography remains an intrinsically inter-disciplinary subject of ever-growing relevance as the pace of global environmental change accelerates. Geographers grapple with the inter-connected nature of the Earth's fundamental geodynamic systems (i.e., lithosphere, hydrosphere, biosphere, and atmosphere) and their impact on, and interaction with, different scales on the human use system. It is by definition both local and global in scale, and geographers bring their unique spatial and analytical skills to bear on these interactions.

So, you are still sitting there and are now no doubt wondering how do I succeed in this subject? How do I gain a fundamental knowledge of physical geography if it is so important? At this stage you probably want me to direct you to the magic 'know it all geography potion'. Sadly, this does not exist.

The only key to success is to read, and to read in an *engaged* manner. That is with the lights on behind your eyes and your brain firmly in gear. Start with this primer and try to make additional notes using the tasks as a starting but not an end point. These notes may end up at worse as a crumpled set of pages on your floor, or at best get filed in a nice shiny, new binder. You may even go as far as to buy a copy of one of the physical geography textbooks available and display it proudly on your shelves. But have you ever heard of the 'psychological value of unused information?' People tend to buy self-help books but never read them but feel better for having them – well that is the concept. It applies here, having that new shiny binder, an electronic copy of this primer, or a textbook might make you feel better, but in truth will not improve your understanding or grade. You need to engage with those notes and books.

Start by reading the relevant chapter of the primer, what do you understand and what has passed you by? What interests you and what left you feeling cold? Look at the suggested reading, does any of the items interest you? Now draw up a prioritised list of things to follow up on based on the answers to these questions. May be at the top are the items you do not understand, or a topic that interested you, maybe it is to read one of the suggested papers, or maybe it is to simply spend half an hour on the Internet getting some specific examples, facts, and illustrations. Whatever it is, add to your notes and this primer by further research. If you do not understand stuff, then be proactive; do not sit there worrying about it, seek help from your peers or instructor. If approached right, geography is and should always be fun.

Contents

Part I
Earth History and Climate

Chapter 1
Geological Time

Entropy

Much of our story deals with entropy, the inevitable process by which everything ends in chaos in the end. We see this in so many different ways from the boulder perched on a cliff top that will ultimately fall, to the differential heat gradients across and within the Earth that drive its essential natural systems. My children do the same to the house; we tidy, clean, and organise and they disrupt, scatter, and bring chaos to that order.

Energy is required to do work and we can think of two types of *free energy* and *heat energy*. One is coherent and structured (free energy), while the other (heat energy) is random and over time heat energy will always prevail. The first law of thermodynamics tells us that the total amount of energy in the universe is preserved, it simply changes from one state to another. The second law of thermodynamics indicates that over time energy becomes more randomly distributed and will flow from areas of high potential (excess) to areas of lower potential to achieve this. As energy moves down these gradients it will be structured or free, and the greater these gradients the greater the flow (flux) of energy and its ability to do work.

Take a pint of beer, and although it seems something of a waste, pour it over the floor. The beer in the glass has energy, potential energy stored by the process of raising the glass, as it flows over the lip that energy is directed into a cascade and has structure, the kinetic energy of the flow. Once it hits the floor and spreads out as a giant aimless puddle it no longer has structure and the energy is randomly distributed. The free energy of the cascade can do work, the aimless puddle cannot. The second law of thermodynamics says that entropy increases over time or put another way chaos always wins.

The more entropy a system has the more random it is, and the less free energy is available to do work. If a huge mountain range is formed, it is inevitable that over time it will erode thereby increasing the entropy of the landscape. In doing so free energy will be driven by gravity as flows of rock, ice, and water. It used to be fashionable to view landscape via an age cycle (i.e., youthful, mature, and old age) a concept championed by William Morris Davis (1850–1934). All that would be left in old age would be a peneplain, a flat surface with relict hills bearing testimony to the mountains of its youth, a landscape high in entropy.

This concept has been largely discredited and replaced by process studies rather than descriptive models, peneplains once thought to be everywhere, are now considered rare. The truth is that outside textbook theory a landscape never attains peneplanation in a single cycle. But there is an element of truth here, however, over geological time scales mountains are eroded, and landscapes are worn away to be submerged below the sea. This is done by a wide variety of processes, controlled by climate and the biosphere, and one life cycle is rarely seen to completion before being interrupted by episodes of rejuvenation such is the dynamic nature of our planet. It provides a frame however to think about the Earth; tectonic forces create mountains, and those mountains are eroded by free energy to render a landscape rich in chaos and disorder!

No Sight of a Beginning, No Sight of an End

One of the key skills of any scientist is to observe what they see without prejudice or bias and record the facts objectively and accurately. What they do with those facts is the next key test, using those observations to make a series of logical inferences, preferably ones that can be tested by more observations is key. What actually tends to happen is people observe and then try to fit those observations into an existing paradigm (theoretical model), or in the competitive world of modern research occasionally claim inferences that are not supported by the evidence! Being led by *observed evidence* is a key skill, irrespective of where that evidence takes you in terms of logical inference even if that is in the face of accepted theory or your own beliefs.

One individual who perhaps personifies this skill is James Hutton (1726–1797) who is described by some as the father of modern geology. A gentleman scientist and Scottish landowner James Hutton was part of an Edinburgh community of Natural Philosophers in the late eighteenth century sometimes referred to as the Scottish Enlightenment. He subscribed to the premise of actualism; the processes that operate today have always operated. James Hutton spent a decade or more as a farmer and observed his fields erode slowly and concluded that such erosion must have always been thus. It was a response to the idea of 'catastrophism' in which Earth history could be explained by a series of catastrophic events, compressed into a short period of time with processes operating unlike those today. The Biblical Flood described in Genesis is perhaps the best example of such a catastrophic event.

It is perhaps worth pausing for a second here to say that the agreed age of the Earth at the time was 9.00 am on the 26th October 4004 BC. If you slept in that morning, you might have missed it. This precise age was published in 1650 by Archbishop James Ussher following a literal reading of the Bible and a consideration of relevant calendars. While we know today that this is not true, we should not mock this work unduly because it was a piece of scholarly work involving a huge polymathic grasp of ancient text and was based on the premise that the age of humans must be approximately co-eval with the age of the Earth. In the eighteenth century the prevailing geological origin story was based around the work of Abraham Gottlob Werner (1749–1817) which started with a universal ocean which slowly receded to reveal the land. Primary rocks were revealed, the oldest and those at the most remote of locations and included all granites. They are also fossil free being formed before life commenced. Transition Rocks came next in this scheme eroded from primitive rocks but with some primitive life and distorted and deformed. This deformation was due to the collapse of underground caverns according to Werner. Secondary Rocks were more recent stratified rocks and were normally horizontally bedded. The final type of rocks in Werner's sequence were Alluvial formed by volcanoes and floods. The Christian community were more or less ok with this scheme since it did not challenge Ussher's date and fitted with Noah's Flood and/or the original waters of creation.

It is with this backdrop that Hutton gave a series of four lectures in 1785 in Edinburgh in which he outlined his theory of the Earth and its great antiquity. How could it be anything but ancient if the processes of erosion and deposition we observe today are so slow? His ideas were received with scepticism at the time, but over the next few years he won over some ardent supporters one of which was John Playfair who would do more than anyone to promote Hutton's work over the coming years.

A key piece of evidence in convincing these supporters was a visit to Siccar Point on the Berwickshire Coast (Fig. 1.1). Here Hutton pointed out to his acolytes the immensity of geological time. The vertical beds at the base of the cliff must have once been laid down horizontally he argued, before being folded and raised in the

Fig. 1.1 The angular unconformity at Siccar Point on the coast of Berwickshire in Scotland. (From an original drawing by Hilary Foxwell)

heart of a mountain, this mountain was then subsequently levelled by erosion before new sediment was laid on top. If you do not countenance any process, you cannot observe today, then this must surely have taken an eternity to occur. What Hutton had found was an angular uncomformity, a break in geological time of huge proportions. These observations challenged not only Hutton's own religious beliefs but also the early ideas in geology most of which resolved around Biblical arguments. Hutton's thesis was challenged by those who clung to a more conventional and Biblical thesis, one of which Richard Kirwan essentially accused him of being an atheist and blasphemer which were strong words at the time. Despite being unwell Hutton resolved to expand his 1785 paper and in 1795 *The Theory of the Earth* was published. Unfortunately for whatever reason, illness, or an inability to articulate his ideas clearly, the book has gone down in history as one of the most poorly written geological texts of all time. It was left to the works of John Playfair to champion Hutton's cause following Hutton's death in 1797.

In the following years geology proceeded rapidly at least in the UK and the basic framework of the geological column was established with its classic periods Cambrian, Ordovician, Silurian and Devonian. The names reflect locations in the British Isles (Celtic Tribes mainly) where thicknesses of rock belonging to each interval were defined. What is important to note is that this was done without any recourse to absolute dating. Today we talk about geological time in millions of years based on a range of radiometric dating methods that use the principle of time-dependent radioactive decay to date things. These absolute methods contrast with the more common process of dating by relative age and/or relative position. Much of the geological record was established, and accurately established, without recourse to absolute dates. Basic stratigraphic principals were used; what lies at the bottom of the cliff, what layers are at the top and how can we correlate or link those layers to others via rock type or more commonly the fossils they contain? A geological period, or its component sub-divisions, are now defined by international agreement and a proverbial 'Golden Spike' is driven into a type section somewhere in the world to define the boundary between one unit of time and the next.

One of the striking aspects of the geological column is how little we actually know in detail. The three big eras Cenozoic, Mesozoic and Palaeozoic collectively known as the Phanerozoic Eon account for just 570 million years of the Earth's 4.6 billion life history. The rest is split as the Proterozoic, Archean and Hadean referred to loosely as the Precambrian. The reason for this lies in a remarkable evolutionary event at the start of the Cambrian, known as the Cambrian Explosion, when there was a huge radiation in complex, multicellular life with hard parts that got preserved, things like shells and bones. These fossils allow us to resolve the chronological order of the rock record. While complex life existed before this time it is not well preserved. Plate tectonics also plays a part here. The constant churn of rocks at the Earth's surface means that older rocks are not well preserved and when they do outcrop at the surface, they have complex histories with multiple events superimposed on top of each other.

- Task 1.1: Chose any geological time period (e.g., Jurassic, Triassic, Cambrian etc.) and find some key facts about it. How long did it last, where are the type sections which mark its start and finish?

The geological record is imperfect, more like a piece of glamorous lace than a solid garment, full of provocative holes. An angular unconformity is one type of hole, but there are many more. The problem is that rocks are only really preserved in sedimentary and volcanic basins. Imagine a flamboyant pastry chef with a large cake tin at which they throw ingredients. Some end-up in the tin, others miss, and a mess is created around. The tin is placed in the oven and the lucky ingredients end up in the final cake, the rest is lost as the table is washed down. So, it is with the rock record. If sediment is laid down in a basin it has a chance of being preserved, if it does not then the chances are that it will be lost to erosion. Think for a moment of Hutton's unconformity; it preserves two rock sequences deposited in two separate marine basins. Nothing is preserved from the period of mountain building and erosion between these. For a long time, the origin of these sedimentary basins was unclear, referred to by such vague terms as geosynclines, but plate tectonic changed that by giving a mechanism by which sedimentary basins could be created and closed. It is one of the reasons why plate tectonics has transformed the discipline of geology, a transformation that has occurred in the last 40 years.

Uniformitarianism or the key to the past is the present is one of the fundamental geological principles. The laws of physics have not changed as far as we know so the same process must have always operated as they do today. Observe the process and the product created and we can infer the process from the same product we see in the geological record. The geological history of Earth has been built up by countless geologists and geographers applying this principle.

Earth's Origin Story

Comic book superheroes all have origin stories and since there is no bigger character than the Earth perhaps it deserves its own origin story? While there are many spiritual and religious versions of this story, our focus here is on the geological origin of our planet. This story is based on a judicious application of uniformitarianism and careful decoding of the evidence left in the rock record. It is a story of mountain building, of changing sea level, climate change and the evolution of life.

The Earth is an average planet orbiting an average star, not to big, not to small. On a Hertzsprung-Russell diagram our Sun is right bang in the middle sitting on the main sequence (Fig. 1.2). It is middle aged and will eventually become a Red Giant expanding to evaporate the Earth and the inner planets of solar system something that will in all likelihood happen in about another 4 or 5 billion years' time. Don't panic you have time to put the kettle on and finish this book.

About 4.6 billion years ago the Earth came into existence by the agglomeration of rock and gas orbiting the Sun. At first it would have been a giant molten ball of

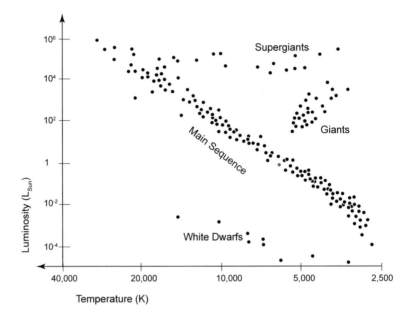

Fig. 1.2 Hertzsprung-Russell diagram, the Sun has a luminosity of 1 and lies in the main sequence. (Modified from images provided online by Richard Pogge, Ohio State University)

matter heated by continual bombardment by smaller planetary bodies. Over time the heavier elements, such as iron and nickel, sank to form a solid core. Around this core a molten layer exists to this day. The combination of this solid and molten core combined with the Earth's axial spin generated a magnetic field which also continues to this day. This magnetic field began to protect the surface from damaging charged particles of solar wind emitted from the Sun. Surrounding this two-layer core is the mantle a three-thousand-kilometre-thick layer of semi-molten rock, gas, and water. The mantle is not entirely molten however, but at high temperatures and pressures rock can flow and it can also change state easily.

The surface of the Earth is warmed gently by heat from the ground, less gently in volcanic regions such as Iceland or Yellowstone where molten rock is close to the surface. This geothermal heat flux is caused not only by the residual heat from the molten core, but also by the radioactive decay of such isotopes as potassium-40, uranium-238, uranium-235, and thorium-232. The geothermal gradient is about 25–30 °C/km and the upper mantle has a temperature of around 1900 °K and the inner-core 7000 °K which is quite hot! The pressure in the core is about 3.6 million times that at the surface. Now over time radioactive isotopes are consumed and 4.6 billion years ago there would have been more of them, and the temperature of the core was consequently twice that is today around 3 billion years ago. The result would have been a stronger temperature gradient core to surface. The Earth's internal heat engine is driven by this thermal gradient and the stronger it is the more vigorous that engine.

- Task 1.2: What is the average geothermal heat flux in Britain or any other country you chose? How does this compare to somewhere like Iceland? A web search should reveal this information.

A pan of tomato soup on the stove bubbles away via a series of internal convection cells. Warm soup rises over the heat to the surface, as it become less dense, and moves to the sides of the pan where it cools becoming denser causing it to sink. Now the Earth's interior also needs to release heat and heat flows along the temperature gradient towards the cold surface via the mantle just as in the pan of soup. Unlike soup the mantle is much more viscous; in fact, it is really a solid that moves slowly. However, the principle remains, heat has to get from the core to the surface. This happens in a variety of different ways, some continuous, some more episodic. If we go back to early molten Earth lighter elements would have naturally surfaced and cooled like a skin on a rise pudding. This early crust was moved over the surface of the planet by the rising heat from the interior. Where gaps occurred in the crust new volcanic material emerged, along with increased heat flow. Here new crust was formed. If you create new crust, then you must compress crust in other areas to make space for it assuming the sphere stays the same size. In areas of compression colder denser crust would sink and return to the mantle helping to cool it. Effectively the churn of the crust and mantle help moves heat from the interior to the surface just as the convection currents do in the soup. The processes involved in this churn within the early Earth are still poorly understood but an area of active research.

Overtime this process resulted in growth and thickening of the crust. This early form of plate tectonics would have built stable areas of crust (cratons), which collided, merged, and split growing progressively as they did. At this point they would have probably been separated by the earliest of oceans. One famous geologist once described this time as 'cornflake-tectonics' with more milk than cereal. With time cratons merged to form larger areas (shields) and the first true continents made of lighter minerals came into existence. At some point the processes of global tectonics began to resemble that we see today, the classic plate tectonic model which we will introduce in the next chapter. This tectonic churn is a continuous way of releasing heat from the Earth's interior.

This continuous process contrasts with a more episodic one involving super-plumes. Super-plumes are unstable episodes in which large masses of molten rock rise from the interface between the mantle and the outer core. They emerge at the surface and give rise to extensive volcanic fields and are not usually associated with conventional plate boundaries although they can give rise to boundaries over time. Today we have modest volcanic plumes such as that beneath the Hawaiian Islands in the Pacific, which are located in the centre of a plate. There is also another modest plume beneath Iceland located on a plate boundary. Super-plumes are much larger in size than these examples and are the product of instability at the mantle-core interface. They have played an important part in Earth's history and in controlling biodiversity.

Today we have an Earth with a solid core, a liquid sheath around and a thick mantle beneath a thin crust known as the lithosphere. This crust is made up of either thin ocean rocks which originate in the mantle, or thicker continental rocks made of lighter volcanic rocks and fragments created by eroding those rocks. Both types of crust rest on a more ductile layer close to the surface of the mantle called the asthenosphere and are split into a series of plates. These plates move with respect to one another at about the same speed at which your fingernails grow. This movement is due to upwelling of molten rock; the pull caused by sinking of colder denser crust; and collectively helps release heat from the Earth's interior.

This classic plate tectonic model has operated throughout the last billion or so years of Earth history during which time the continents have done a slow shuffle. Assembling to form a giant supercontinent Pangaea which reached its zenith about 335 million years ago, before breaking up to give the continental distribution we see today. Geologists project that in 250 million years' time the continents will once again be assembled into a giant supercontinent, something that some refer to as Pangaea Ultima. If we look into the early rock record some see evidence of earlier supercontinents that pre-date Pangaea, others do not. There is some consensus about a supercontinent Rodina in the late Neoproterozoic, but prior to that there is uncertainty.

The question centres around whether the global tectonic model we see today has always operated or not. One widely used analogy is to think of a cockerel, to look at it how would we know it was once an egg? This is where the principle of uniformitarianism (key to the past is the present) begins to break down. What came before the global tectonic model we see today? We will explore this a little in the next chapter, but the key point for now is that Earth's surface is constantly being refreshed. Many of the moons and planets of our solar system show evidence of intense percussion from asteroids between 4 and 3.8 billion years ago caused perhaps by shifts in the orbits of Jupiter and Saturn. The absence of plate tectonics means that these scars have not been removed. Thankfully, this period was relatively short-lived, and most asteroids are located between Jupiter and Mars the building blocks perhaps of a planet that never was.

We can contrast the Earth's surface with that of the moon which is pitted by impact craters and has remained unchanged for millions of years. The supercontinental cycles of the Phanerozoic at least have kept the Earth's surface fresh, mountains have been created and eroded. In fact, we can see this as the application of entropy. Mountains are great stores of potential energy that is released as they are stripped back and worn-down by erosion. Erosion is an agent of entropy.

Further Reading[1]

Bryson, B. (2016). *A short history of nearly everything*. Transworld Publishers.

---

[1] This chapter is mainly background but if you are interested there are some good popular science accounts which cover some of this material well. I would recommend David Christian's (2018) *Origin Story* published by Allen Lane. It is a popular science book which as the title tells gives a

Christian, D. (2018). *Origin story: A big history of everything*. Hachette UK.

Gould, S. J. (1987). *Time's arrow, time's cycle: Myth and metaphor in the discovery of geological time*. Harvard University Press.

Hallam, A. (1990). *Great geological controversies*. Oxford University Press.

Repcheck, J. (2010). *The man who found time: James Hutton and the discovery of the Earth's antiquity*. Simon & Schuster.

Smithson, P., Addison, K., & Atkinson, K. (2013). *Fundamentals of the physical environment* (4th ed.). Routledge.

---

scientific version of the Earth's origin story and is well-written, informative, and wide-ranging. Another popular account worth dipping into is Bill Bryson's *A Short History of Nearly Everything* published in 2016 by Transworld Publishers. A more serious read is the book by Hallam (1990) *Great Geological Controversies* published by Oxford Science Publications. This book recounts the early debates in geology including the tussle between uniformitarianism and catastrophism. In particular, Chaps. 2 and 6 give insight into deep time and the origins of the plate tectonic paradigm. Sticking with popular science Jack Repcheck's *The Man who found Time* (2003) published by Simon & Schuster gives an account of James Hutton's contribution. There is one other popular account of deep time that might be of interest written by Stephen Jay Gould (1991) entitled *Time's Arrow, Time's Cycle: Myth and Metaphor in the Discovery of Geological*. A reproduction of Hutton's famous book *Theory of the Earth* was published by Classic Books International in 2010 but can also be read and downloaded from: www.gutenberg.org/ebooks/12861. Most textbooks carry a short section on geological time and my preferred one Smithson et al. (2013) *Fundamentals of the Physical Environment* is no exception.

Chapter 2
Earth's Internal Heat Engine

As we saw in the last chapter plate tectonics is driven by the temperature gradient between the Earth's surface and its core. It is driven by the free energy that flows along that gradient and its vigour is determined by that gradient. It is worth spending a bit of time looking at the current tectonic regime that operates on the Earth today, because not only does it give us the framework to interpret past geological events, but it also defines the macro-scale geomorphology (surface morphology) of our planet. Before doing so let us spend a few moments thinking about how the concept emerged, which although now pervasive is relatively new.

Paradigms and Shifting Science

A paradigm is a model or framework by which something is understood and through which evidence can be interpreted. Like most people scientist like to conform, be liked, be popular and be acclaimed, they are after all only human. There are also few truly brilliant ones and they often make their mark outside their home discipline by looking at existing evidence with new perspective. Science proceeds a bit like a car on a rollercoaster via a series of adrenalin fuelled revolutions (paradigm shifts), between which there is calm.

Scientist naturally fit, where possible, their data into the scientific ideas of the day, often those that have been *'bred into the bone'*. It is natural for these ideas to *'out in the flesh'* especially in one's home discipline. The human desire to conform and be liked by one's peers reinforces this. Overtime an existing paradigm can become difficult to maintain, however, in the light of new data although scientists within that discipline may be blind to the limitations. That is where new perspective (and data) helps; an outsider can often see what they cannot and more to the point is

M. R. Bennett, *Our Dynamic Earth: A Primer*,
https://doi.org/10.1007/978-3-030-90351-0_2

often more prepared to say so. Afterall if their ideas are mauled, they can retreat to their own home discipline. New ideas do not conform, and therefore they usually become controversial. Some are dismissed, and perhaps rightly so, others linger until better data comes to light, few are accepted outright without a fight. Whether the revolution happens soon after a new paradigm emerges or happens after a period of many years it is always quick and is followed by a slew of confirmatory research papers as people jump onto the new and suddenly fashionable idea.

This holds true for plate tectonics, or as it was first known continental drift, and was pioneered not by a geologist but by a meteorologist. His name was Alfred Wegner (1881–1930). One apocryphal tale has Alfred watching icebergs off the coast of Greenland when he had his big idea. At the time (1906) he was serving as a meteorologist to an artic expedition. More likely, and as stated in his own writings, is that like many before him he looked at the geographical fit between Africa and South America and begin to wonder especially as he read a report on their palaeontological similarity. He aired his theory for the first time in a lecture in 1912 and his book *Die Entstehung der Kontinentr und Ozeane* (Translates to: *The emergence of the continent and oceans*) was published in 1915, with later edition in 1920, 1922 and 1929. The third edition was translated into English and the term continental drift became widespread despite the fact that Wagner's actual description was 'Die Verschiebung der Kontinete' which translates as continental displacement. It is worth stressing here that the idea of Pangea, a giant supercontinent, was central to Wegner's ideas.

Despite amassing large bodies of data to support the idea that the world's continents once formed a giant landmass (Pangea) the geological community largely rejected the idea (Fig. 2.1). The problem was Wegner was unable to present a viable mechanism for how continents moved. The prevailing idea was that as the Earth cooled it contracted and this was what caused the deformation visible in mountain ranges around the World. So, the matter rested until after the Second World War.

The Second World War saw a huge surge in technical know-how that was applied for peaceful purposes once the war was over. The geophysical techniques developed at this time were to transform our understanding of our planet. The first real contribution was the development of palaeomagnetism made possible by a highly sensitive magnetometer. Iron-rich igneous rocks acquire a magnetism from the Earth's magnetic field which becomes set in the rock's minerals as they cool below the Curie Point. Think of lots of little compasses in the molten rock that align to the poles, once that rock cools that magnetic signal becomes locked into the rock crystals. If the poles, then move you can establish a difference between the rock's palaeomagnetism and that of the current polar position.

The magnetic poles wander, but if the continent also moves then we can show the relative position of both in the past given enough samples. With the development of radiometric dating, the ability to date rock, and attain their palaeomagnetic signal became possible. The discovery of polar reversals quickly followed; at regular intervals throughout geological time the poles have reversed. These reversals are preserved in the palaeomagnetic signals of lavas. The final piece of the puzzle was rapid mapping of the ocean floor using echo-sounding equipment. These maps

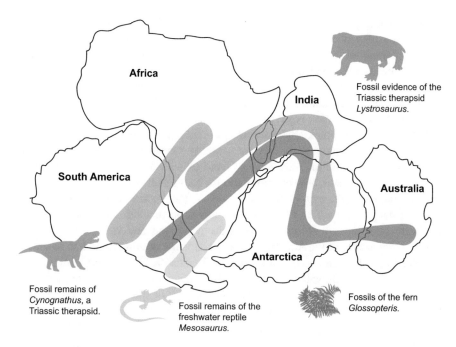

Fig. 2.1 Palaeontological evidence for Continental Drift as proposed by Alfred Wegner. (Palaeontological distribution and continental outlines from Osvaldocangaspadilla, with silhouettes from phylopic.org)

established the presence of the rugged Mid-Atlantic Ridge and the smoother East Pacific Rise. They also established the presence of deep ocean trenches and while dredging from survey ships found no rocks older than the Cretaceous on the ocean floor.

Putting all these elements together, Harry Hess (1906–1969), published the seafloor spreading hypothesis in 1962. Hess suggested that the ocean floor was created at mid-ocean ridges and moved towards ocean trenches where they descended into the mantle. Confirmation was to come a few years later with the recognition of palaeomagnetic stripes either side of the mid-ocean ridges. Finally, Tuzo Wilson articulated the new paradigm of plate tectonics in 1965, since when it has revolutionised our understanding of planet Earth.

Plate Tectonics

Let us explore this paradigm in more detail and focus for a moment on the Earth as it is today. The Earth's surface is split into a series of giant plates made up of a combination of ocean and continental crust. The former is derived from the mantle, the rocks are depleted in silica and described therefore as basic. Ocean crust is also thin around 5 km thick. In contrast, continental crust is thicker (30–50 km) and more diverse in composition with a range of sedimentary, metamorphic, and igneous rocks. Sedimentary rocks are the eroded products of other rocks, while igneous rocks are produced by volcanoes and the by the intrusion of molten rock (magma) into existing, country, rock. Metamorphic rocks are rocks changed by heat and pressure either during mountain building or the intrusion of igneous rocks. We can conceptualise these three rock types as a simple rock cycle (Fig. 2.2).

We can detect the boundaries between different plates through the distribution of earthquake and volcanoes since they are naturally concentrated at plate boundaries. Today we can measure the movement of these plates via satellite-based telemetry; literally we can measure how two points on the Earth's surface change position over time. We recognise three types of plate boundary, constructive, destructive, and conservative margins (Fig. 2.3). Constructive margins occur where two plates are moving apart. New crust is being created as the gap opens and we find this type of margin in the centre of the Atlantic, Indian, and Pacific oceans today. The boundary is marked by a continuous rift valley, bounded either side by a chain of mountains and in the floor of that valley there are active volcanoes and lots of vertical intrusions (dykes) of igneous rock. The geomorphology of these sea floor spreading centres is determined by the speed of spreading. If the plates are moving apart fast

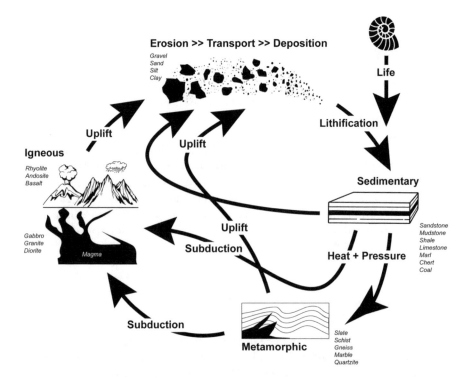

Fig. 2.2 Rock Cycle showing the relationship of their three main types of rock to one another and the key role of plate tectonics in terms of spreading, subduction, and uplift

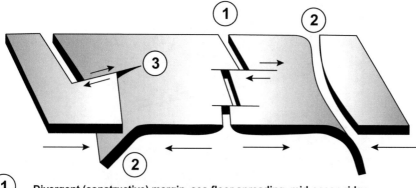

1. Divergent (constructive) margin, sea-floor spreading, mid-ocean ridge

2. Convergent (destructive) margin, subduction zone,
   volcanic arc or continental margin orogen, ocean trench

3. Conservative margin, transform fault

Fig. 2.3 Main plate tectonic boundaries and the various names used

then the topography is smoother and the axial rift less rugged and deep. Fast spreading is associated with a higher heat flux and supply of molten rock so the whole plate boundary is more buoyant and has a greater volume. In contrast if the speed of spreading is slow, as it is in the Atlantic, then the boundary is less buoyant and the total volume smaller. The lack of heat means that the rocks are more brittle, and the axial rift is usually deeper and more rugged. Sea floor spreading centres have been described as the longest chain of mountains and volcanoes on the Earth's surface, it is just that we cannot see them.

- Task 2.1: Find out about the Mid-Atlantic Ocean Ridge how does it compare to the East Pacific Rise in terms of length, topography, and activity?

In a few places around the world, such as in East Africa, divergent margins occur above sea level. Here continental crust is splitting apart. Overtime it will if the two plates continue to move apart and an ocean will be created. This is the first part of what is often referred to as a Wilson Cycle, the opening part of a cycle (Fig. 2.4).

Elsewhere on Earth plates must collide if they are moving apart at other locations and the diameter of the planet is constant. At these locations we have subduction where one plate descends under another. Subduction zones (also termed convergent or collision margins) come in two types. We have a Continental Margin Orogen, where ocean plate meets continental crust. Here the denser ocean plate descends beneath the continental crust which causes compression in the leading edge of the continental plate. Sediment and water are dragged into the subduction zone with the descending plate and the friction creates heat. Molten rock results which is rich in silica, in contrast to the ocean plate which is depleted in silica. This molten rock works its way to the surface and creates a line of volcanoes in the compressed

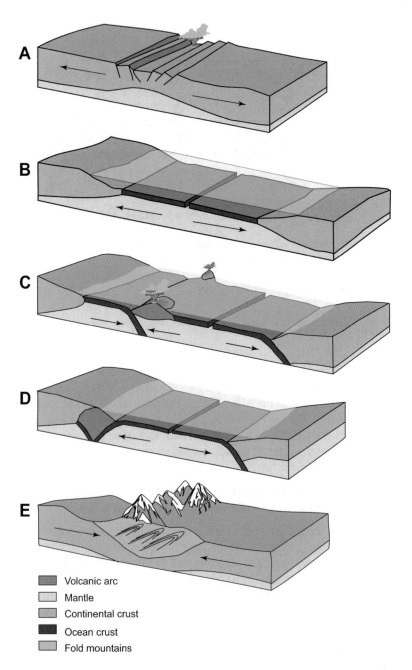

Volcanic arc
Mantle
Continental crust
Ocean crust
Fold mountains

Fig. 2.4 The Wilson Cycle. The switch from divergence to convergence is influenced by the inter-play of plates moving elsewhere on the Earth's surface

leading edge of the continental plate. The Andes is a classic example of this today, mountains and volcanoes occur along the leading edge of the South American Plate caused by of the Pacific Ocean Plate descending beneath it. This is the classic sub-duction zone. Things to note are the ocean trench which marks the line of collision and the fact that the mountains/volcanoes are set back from this line by a distance that depends on the angle with which the ocean plate descends.

The alternative type of subduction zone is where two ocean plates collide. The plate that is cooler, and therefore denser, will descend below the other one. Plate temperature is a function of the distance from the spreading centre at which the plate was created, greater the distance the more opportunity for the crust to cool. Again, ocean water and sea floor sediment are dragged down into the subduction zone where it is melted by friction. The molten rock rises to the surface through the leading edge of the overlying plate to erupt first as sub-ocean volcanoes that emerge in time to create volcanic islands. These volcanic island arcs are common today on the western side of the Pacific Ocean and form part of what is known as the Pacific Ring of Fire. They describe arcs on the surface of the planet since the planet is a sphere. Subduction and the consumption of ocean plate is the second half of the Wilson Cycle and will ultimately lead to orogenesis, which is just a fancy way of saying mountain building (Fig. 2.4).

- Task 2.2: Find some examples of volcanic island arcs, where are they located, what are their geographical characteristics?

The third type of plate boundary is a transform boundary where plate of whatever type is neither created nor destroyed plates just slide past one another. The San Andreas Fault in California is the classic and much cited example.

In summary, our planet's surface is composed of plates which move relative to one another. As they do so they create and close sedimentary basins in which rocks first accumulate before being deformed and perhaps preserved. The macro-scale topography of our planet is controlled by these plate boundaries. There is a reason that we do not have extremes of elevation of depression on Earth and that there are basically just two topographic levels, that of the ocean floor and that just above sea level. The fact that the ocean crust is lower than continental crust, and therefore allows oceans to form, is not just a function of crustal thickness but also a function of buoyancy. To understand this, we need to explore the concept of isostasy.

Isostasy

Topography on the Earth's surface is a product of several things, variations in crustal density, active and/or passive uplift, erosion, and sedimentation. Crustal density depends on rock type and temperature. The complicating factor here is isostasy and as a result topographic elevation is not just a simple matter of squeezing or faulting rocks upwards to create mountains.

Archimedes' Law states that when you place a block of wood in a bathtub full of water, the block sinks until the mass of the water displaced by the block is equal to the mass of the whole block. Density is mass in a unit volume, so change the mass or change the volume and you change the density. Continental crust has an average density between 2.6 and 2.7 g cm³. In contrast, ocean crust has a basalt-like composition, with a density closer to 3 g cm³. Temperature come into play via volume; warm things expands so a unit length of ocean plate has a greater volume for the same mass when warm than cold. This is why ocean plate becomes denser with distance from the sea floor spreading site at which it was created.

Now you might rightly say that Earth's plates are not floating in water, which is true, but they rest on the more ductile asthenosphere which acts in a similar way although with a much greater density and viscosity. There are two alternative models for isostasy both of which are correct. One was developed by Calcutta's archbishop John Pratt (1809–1871), and the other by George Airy (1801–1892), Royal Astronomer and Mathematician. Pratt suggested that blocks of terrain reached a level of buoyancy effectively dependent on their density. So less dense crustal areas would stick up more than denser ones (Fig. 2.5). Alternatively, Airy suggested that in a situation where density was constant the amount a block would protrude would depend on its size. So, in this model mountains must have deep roots, which they do. Both positions are correct, and Pratt's model applies best in the oceans, while Airy's model best describes the situation on the continents. The essential point here

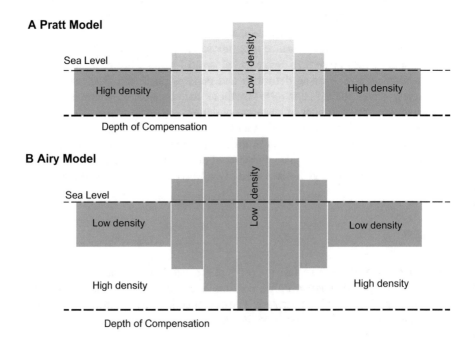

Fig. 2.5 Pratt and Airy models of isostasy

is that if you change the density of a unit area of the Earth's surface you change its topographic elevation.

One of the implications of the Pratt Model, and the role of density, is for sea level. We have two types of sea level eustatic, which means global sea level, and the second is relative sea level which is a local or regional phenomenon. If we have a seafloor spreading site that is particularly hot and active the ocean crust either side will be warmer and therefore less dense. It will as a result be more buoyant and consequently displace more water. So high eustatic, or global, sea levels are usually associated with lots of seafloor spreading. During the break-up of Pangea in the Mesozoic there was lots of seafloor spreading and sea levels were correspondingly high. The chalk of the White Cliffs of Dover was laid down during this time in a warm epicontinental sea. Epicontinental sea simply means a sea on a flooded continental surface.

The Airy Model has implications for topographic elevation on the continents. If over time a mountain range is eroded its mass is reduced. Put another way the block with our mountains on it is reduced in elevation. As a result, it will rebound as the load is removed and be uplifted. Similarly, the eroded sediment ends up somewhere, for example a large coastal delta at the coast, it will depress the crust by increasing the load at that location. Subsidence allows more sediment to be accommodated in the delta. The point here is that topographic elevation is not just about folding and thrusting rock upwards but also about shifting the load from one location to another causing adjustments in elevation.

There is one special variant on the above that is caused by the growth and decay of large ice sheets. When we talk about big, I mean things the same size or larger than the ice sheets in Greenland or over Antarctica today. An ice sheet is a load and over time it will depress the crust beneath it, as that ice sheet melts the crust will rebound. The point to remember here is that the Earth's crust is slow to respond so the depression and rebound is often out of phase with the growth and decay of the ice sheet. This has big implications for relative sea level in some areas of the World once glaciated like Scotland, Norway, or parts of North America.

Supercontinents Again

We introduced the idea of supercontinents as part of the Earth's origin story in the previous chapter. Pangaea was central to the original idea of continental drift and first its assembly and then dis-assembly dominates much of the geological history of the Phanerozoic. But what about earlier events?

This is tied up with the question of whether plate tectonics has always operated as it does today or whether the current tectonic model is just a phase in our planet's life cycle. You might ask why this question matters to a geographer? There are various responses to this, the main one being that physical geography and geology are two symbiotic disciplines and have much to gain from one another. The other answer is that understanding our own planet and its history will help us in the search for exoplanets and life elsewhere in the solar system. We are not necessarily looking for planets just like the Earth or more to the point at the same point in its life cycle. So, let us take a minute to explore this a bit further.

One approach to understand the alternatives to plate tectonics is to look at large rock based planetary bodies in our solar system. We can measure the presence of active tectonic, past, or present, by looking at the degree to which the surface has been refreshed by tectonics. We have seen how the Earth's surface is refreshed by plate tectonics and the evidence of ancient impact craters removed. There is a lot of debate and uncertainty around this question, but one attractive model defines to end members (Fig. 2.6).

At one end are planets and moons that are molten. In this case heat transfer is direct via a surface magma pool. Earth was once like this in the early days. At the other end of the spectrum, we can identified planets that have a strong, thick, and

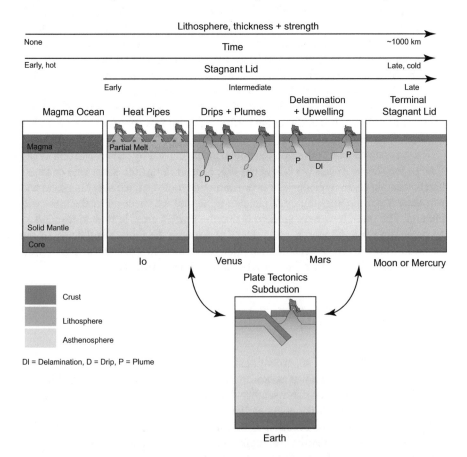

Fig. 2.6 Alternative tectonic or non-tectonic settings within our solar system for large silicate bodies, like the Earth. (Modified from: Stern, R.J., Gerya, T. and Tackley, P.J., 2018. Stagnant lid tectonics: Perspectives from silicate planets, dwarf planets, large moons, and large asteroids. *Geoscience Frontiers*, 9, 103–119)

stagnant crust acting as a stable lid. This is the terminal stage in the model and awaits all planets and moons in due course, just as most probably started with molten surfaces. Between these two extremes are various tectonic models by which heat can be liberated. At the hot end is a simple system of semi-permanent pipes sitting above a molten sub-layer. This is an effective way of venting planetary heat and has been observed on Jupiter's moon Io. Venus and Mars both have a stagnant lid with heat and molten rock supplied to the surface by molten plumes originating in the mantle. The returning cooling flow occurs either by drips, or the delamination of colder volcanic rock from the bottom of the lid. Either of these two models may have evolved into the plate tectonics that we see on Earth certainly during the last billion years or so. The question is when did this change take place?

Some have argued that plate tectonics is a recent phenomenon within the last 700 million years or so. Others have argued that it started early in the Archean. The question in part resolves around whether you believe some of the early supercontinent reconstructions, since the assembly and dispersal of supercontinents is seen as evidence of plate tectonics. There is also a quiet period in Earth's history referred to colourfully as the Boring Billion, or the Dullest Time in Earth's History. This period spanned 1.8–0.8 billion years ago (middle Proterozoic). It is characterized by tectonic stability, climatic stasis, and stalled biological evolution. Some believe this corresponds to a period of time when a stagnant lid may have existed in the form of one large stable supercontinent or series of super-cratons. Others suggest that plate tectonics simply stalled during this time. It is interesting that it restarted with the assembly of the supercontinent Rodina, a near-global glaciation and a bit later by the Cambrian explosion which saw and an evolutionary explosion.

It has been argued that the Boring Billions may have been a slingshot for the evolution of complex life. Reduced nutrient supply during a stable tectonic regime may have increased competition leading to sequential stepwise evolution and diversification of complex eukaryotes. In due course this triggered evolutionary pathways that made possible the later rise of micro-metazoans and their macroscopic counterparts.

Further Reading[1]

Hallam, A. (1990). *Great geological controversies*. Oxford University Press.
Smithson, P., Addison, K., & Atkinson, K. (2013). *Fundamentals of the physical environment* (4th ed.). Routledge.
Wilson, J. T. (1965). A new class of faults and their bearing on continental drift. *Nature, 207*, 343–347.

---

[1] Plate tectonic is a key earth system and you need to be familiar with different types of plate boundary and their locations. Most textbooks have sections which provide a systematic review of plate tectonics and my preferred one Smithson *et al.* (2013) *Fundamentals of the Physical Environment* is no exception. Hallam (1990) *Great Geological Controversies* reviews the rise of the plate tectonics from the ideas of Wegner.

Chapter 3
Volcanoes

Volcanoes have appeal, glamour, and style, but in truth they have a relatively modest impact on the surface morphology our planet. There spatial distribution is point specific and their regional impact modest. But as we will see in a later chapter, they have a profound impact on the Earth's long-term climate regulation via their control on atmospheric carbon dioxide.

Volcanoes sell something of a false image and your average volcano does not meet its popular stereotype. Hollywood depicts them as violent, dangerous, and composed of rivers of orange molten rock, yet in truth most volcanoes are hidden on the sea floor or produce clouds of dark ash and gas. Far less glamourous than a river of molten lava! Carbon dioxide is by far the most voluminous product of volcanism, and it is this that impact on the Earth's climate. Do not get me wrong I like volcanoes and have researched aspects of them for several years, but we should perhaps not delay ourselves to long. The first place to start on a quick tour is to consider their spatial distribution, before moving on to review the process that give rise to the morphological diversity, we see around the World today.

Distribution in Space and Time

In the previous chapter we saw that volcanoes tend to concentrate on plate boundaries. In fact, that is one of the ways we can map those boundaries. This makes sense since these fractures in the lithosphere provide a route for molten materials to move to the surface from the mantle. Volcanoes occur predominantly along mid-ocean ridges and subductions zones. At a more regional scale we can define five spatial classes.

The first of these is volcanic arcs (Fig. 3.1). These occur at subduction zones formed by an ocean plate descending below another ocean plate. Wet sediment is dragged from the ocean trench into the subduction zone where it is melted by

M. R. Bennett, *Our Dynamic Earth: A Primer*,
https://doi.org/10.1007/978-3-030-90351-0_3

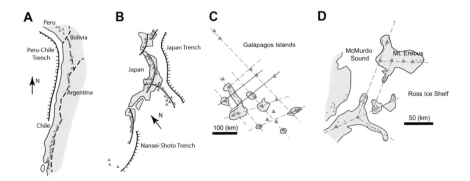

Fig. 3.1 Examples of volcanic distributions. (**a, b**) Chains and arcs. (**c, d**) Examples of clusters showing the importance of fault and join patterns in determining local distributions. (Modified from Clapperton, C.M., 1977. Volcanoes in space and time. *Progress in Physical Geography* 1, 75–411)

friction (i.e., between the two plates) to form magma which rises through the leading edge of the overriding plate to erupt as chains of volcanoes which build volcanic islands. There are approximately 26 volcanic arcs around the World today, most of which are located in the Pacific along its western boundary.

We can then recognise two types of volcanic chain, one terrestrial (Fig. 3.1) and one subaqueous. Terrestrial chains form along subduction zones where ocean plate descends beneath continental plates. The process is the same as with island arcs, but the molten material moves toward the surface through the leading edge of the overriding continental plate (Fig. 3.1). These contrast with the other type of volcanic chain which is hidden from sight below the ocean. Mid-ocean ridges are (i.e., spreading centres) occur where two plates move apart. As they move apart molten material rises to the surface to create volcanoes, in fact a chain of volcanoes marks the rift line between the two plates. The volcanic rock is basic, depleted in silica, and derived directly from the mantle. As it emerges at the surface it cools rapidly to form discrete pillow-like bulbs which stack and squash together under their own weight. The name for such forms is pillow lava (Fig. 3.2).

Volcanic ash is also produced by the sudden quenching of the lava. We talk about ash, but we really mean tiny fragments of volcanic glass, not ash as found in a wood fire or the end of a cigarette. The glass results from the rapid cooling and as the lava is quenched the gas bubbles within it continue to expand and shatter the glass into tiny pieces or volcanic ash sometimes called tephra or tuff. Mid-ocean ridges are rarely exposed above sea level with the exception of Iceland. Here a thermal plume is coincident with a spreading centre and the enhanced volcanism created an island sitting on the spreading centre. Fissure style eruption and low shield volcanoes are the main volcanic forms.

We can then identify a number of volcanic clusters which occur away from plate boundaries (Fig. 3.1). These result from plumes of magma rising from the mantle. The pattern of active and extinct volcanoes tends to reflect the macro-scale

Fig. 3.2 Pillow lava. (Reproduced with permission from Shutterstock, Gudjon E. Olafsson)

joint patterns present. For example, this is rectilinear in the case of the Galapagos Islands and radial in the case of Madeira. Some of these clusters, such on the East African Rift Valley, occur on incipient plate boundaries, a spreading centre in the making, while others do not. Volcanic lines occur where a plate is moving above a stationary plume. This is a bit like a conveyor belt moving above the hotspot; directly beneath the hotspot we will have an active volcano and associated volcanic island emergent above sea level. Behind there will be a line of volcanic islands increasingly inactive and as they cool with movement away from the hotspot. Remember that cooler rock is denser and due to isostasy will tend to sink. Hawaii is the classic example of this (Fig. 3.3); an active volcano at the leading edge of a chain of volcanic islands and submerged seamounts which chart the vector of the plate over the hotspot. A stationary plate will give a cluster of islands and the overall distribution of islands and volcanoes will likely reflect the major fault lines in the plate.

- Task 3.1: Look at the bathymetry of the Hawaiian volcanic chain, a simple web search will suffice. What other volcanic chains are there in the Pacific Ocean?

All in the Eruption

So why do volcanoes erupt? It is a simple question and an easy one; do not get distracted by lots of YouTube videos involving Mentos and soda! As molten rock approaches the surface the amount of rock, and therefore pressure, above decreases. As the pressure decreases the gasses within it begin to exsolve. This process accelerates dramatically as the molten rock emerges at the surface. It is equivalent to taking the cap of a bottle of beer, or champagne if you prefer. As you release the cap,

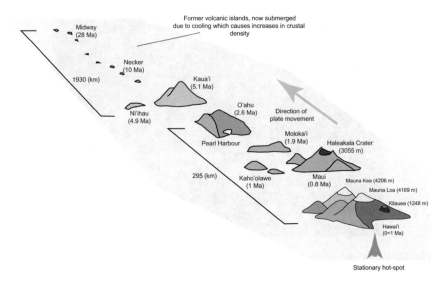

Fig. 3.3 Schematic diagram of the Hawaiian hotspot and volcanic chain. (Modified from Mueller-Dombois, D., 2002. Forest vegetation across the tropical Pacific: a biogeographically complex region with many analogous environments. *Plant Ecology* 163, 155–176)

or cork, you release the pressure under which the liquid was bottled. The first thing is that carbon dioxide within the drink begins to exsolve and bubble rush to the surface. These bubbles expand constrained slightly by the shape of the bottle's neck and the viscosity of the liquid. Gas pressures do not build up. But if our liquid were more viscous then the gas bubbles would not expand as easily and pressure, potentially explosive pressure, would build up within them. That is the key to why volcanoes explode.

Now the viscosity of magma varies with its chemistry which in turn is determined by the source of that magma. Magma derived from the mantle is low in silica, we call it basic magma, and it has a low viscosity as a consequence. Pressures do not build up and consequently the eruption tend to be more effusive than explosive. So basic magmas, those from plumes and at mid-ocean ridges, are not explosive and tend to effuse and bubble to the surface. Lava is the main product. However, magma derived from the melting of sedimentary rocks in subductions zones is usually rich in silica, we call it acidic magma, and acidic magma is viscous. Gas pressures build up in such magmas and eruptions tend to be extremely explosive. As a consequence, the main volcanic product is volcanic glass; think about all that lava being flung into the atmosphere, quenching, and shattering into tiny fragments of glass.

There is one additional element here and that is water. If there is groundwater or shallow surface water the explosivity of an eruption may increase further as this water is turned to steam. A phreatic eruption involves a steam-blast caused by the extreme temperature of the magma (500–1170 °C) and the near-instantaneous evaporation of water. Phreatic eruptions are often associated with the emission of large amounts of volcanic gas (carbon dioxide or hydrogen sulphide). For example, a phreatic eruption on the island of Java in 1979 killed 140 people, most of whom were overcome by poisonous gases.

If an eruption occurs below water, on the sea floor for example, the eruption is also explosive but moderated by the availability of water. These phreatomagmatic eruptions involve the production of large amounts of volcanic ash called Hyaloclastite. This is produced by the rapid quenching and shattering of magma. Hyaloclastite is often found with pillow lavas and creates a significant volume of volcanic material, which is one of the ways that volcanic islands emerge quickly from below water.

Volcanic morphology is determined by a range of factors including: (1) magma chemistry which relates to the magma source: (2) magma volume; (3) the presence or absence of water; and (4) the life-history of the volcano. The last point is important and in general terms we can distinguish between monogenetic and polygenetic volcanoes. The former involves a single, usually short-live, eruption, while the latter involves multiple phases often spread over anything from decades to millennia. In the latter case magma chemistry may change over time as different levels in a magma chamber beneath the volcano are tapped. Big explosive volcanoes have a habit of getting more explosive over time and ultimately destroying themselves to form huge craters known as caldera.

We can think of a continuum from the non-explosive, let's call it effusive, eruption at one end to the extremely explosive eruption at the other. At the effusive end, the magma is basic in composition, and we have an Icelandic or Hawaiian type eruption. Both these types of eruption involve the outpouring of fluid lava building up sheets to form low-angled shields or extensive lava fields and plateaus. The lava is sometimes described as pahoehoe with a smooth, ropey surface texture. The surface may cool and solidify quite quickly but flow may continue via lava tubes and tunnels below the surface. In some cases, lava fields may inflate from within rather than be laid down by a succession of layers. The best examples of shield volcanoes are those of Hawaii. Mauna Loa, for example, rises over 10,000 m from the sea floor (4170 m above sea level) and is over 200 km in diameter. The crest consists of a broad plateau with a shallow crater on top between 4 and 6 km in diameter. This is dwarfed by the Martian shield volcano of Olympus Mons which is over 20 km high and 400 km in diameter at the base.

If the volume of magma is very high, then a lava field of stacked, sub-horizontal flows form. The Columbian River Plateau for example consists of hundreds of individual flows over an area of 130,000 km^2 and has a total thickness of 2000 m. The Deccan Traps covers 500,000 km^2 to a depth of over 2000 m, while Icelandic flows cover 55,000 km^2 to a thickness in excess of 5000 m. The Laki Fissure eruption of 1783 produced over 10 km^3 of lava from a 30 km long fissure over an 8-month period between June 1783 and February 1784. Hydrofluoric acid and sulphur dioxide gas released (circa. 120 million tons) contaminated soil across the island leading to the death of over 50% of Iceland's livestock and the destruction of most crops. The resulting famine killed approximately 25% of the island's human population. Globaly the Laki eruption caused a drop in global temperatures and crop failures throughout Europe and may have also caused droughts in North Africa and India.

If we now move towards the more explosive end of the spectrum, where the magma is more acidic in composition a range of volcanic cones are formed. Amongst them will be the classic strato-volcano cones built from layers of ash and lava. Volcanic ash has a range of forms from tiny pieces of glass, via larger lava fragments (lapilli) through to 'cow-pat' like volcanic bombs. Generally speaking, the larger the fragment size, the closer to the vent it should fall. Figure 3.4 illustrates this idea conceptually; large fragments accumulate close to the vent building up a scoria or cinder cone. This cone will grow until the sides have a slope equal to the internal angle of friction of the ash. The internal angle of friction of a pile of ash, or any dry sediment for that matter, is determined by the packing of individual grains and the degree to which they lock together via surface roughness. If you continue to steepen the slope beyond this angle, then the sediment/ash will fail. Scree or talus slopes reach and maintain this angle as they grow. Volcanic cones are the same (Fig. 3.4). Small cinder cones, often produced by monogenetic eruptions, are quickly eroded both during and after an eruption. They are basically a loose collection of grains. Lava flows may help stabilise the surface and the intrusion of dykes (vertical sheets) and sills (bedding parallel sheets) also add strength to a cone. It is these elements that allow a strato-volcano to develop, with sides that are often close to the internal angle of friction.

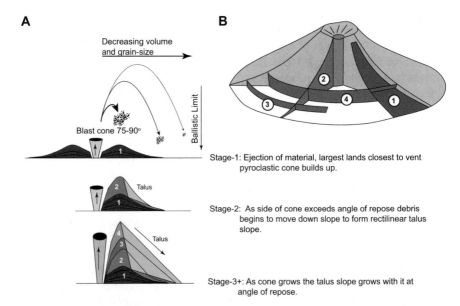

Fig. 3.4 Conceptual model of the formation of a cinder cone and the importance of the evolution of a talus slope in determining the final morphology. The inset shows various volcanic intrusions and surface flows that may help stabilise a stratovolcano (1 = lava flow; 2 = radial dyke; 3 = ring dyke; 4 = cone sheet). (Elements modified and combined from Rittman, A., 1962. *Volcanoes and their activity,* New York. Interscience Publication *and* McGetchin, T.R., Settle, M. and Chouet, B.A., 1974. Cinder cone growth modelled after northeast crater, Mount Etna, Sicily. *Journal of Geophysical research* 79, 3257–3272)

Ash does not just fall from the sky but may also be directed horizontally by blasts of gas. These pyroclastic flows that may be either hot or cold; hot flows (nuée ardentes) weld to form thick layers of ash and can be deadly to all in their path (Fig. 3.5). You have probably seen the images from Pompei with the bodies encased in hot ash, this is what nuée ardentes can do. As the acidity of the magma increases so does the explosivity and move progressively from a Strombolian eruption (Table 3.1) which are moderately explosive with a mix of acidic and basic lava and build up cinder and lava cones to more explosive forms. As the magma gets more acidic, we move up the eruptive scale to Vulcanian associated with large ash cones and block cones with explosive craters. Beyond this we have Vesuvian, Plinian eruptions followed with Peleean as the most explosive with large-scale ash production, although little often remains of the volcanic cone due to explosivity of the eruption. At the extreme end of the spectrum, we have a Karatoian eruption (Table 3.1).

The 1883 eruption of Krakatoa (Indonesia) in the Sunda Strait began on the afternoon of Sunday, 26 August 1883 and peaked on the following morning (Monday, 27 August 1883), when over 70% of the island of Krakatoa and its surrounding archipelago were destroyed by the eruption as the volcano collapsed into a caldera. There

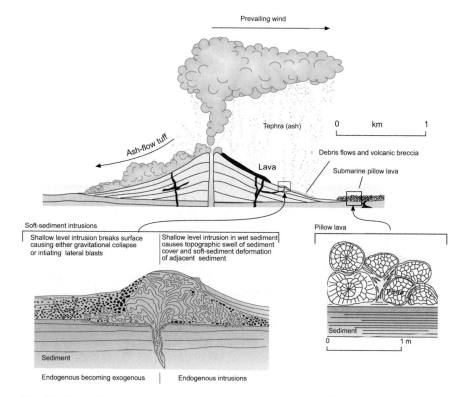

Fig. 3.5 Conceptual model of a volcanic arc eruption involving acidic magma. (Modified from Howells, M.F., Reedman, A.J., Campbell, S.D.G., 1991. *Ordovician (Caradoc) marginal basin volcanism in Snowdonia (NW Wales)*. Special Monograph of the British Geological Survey, HMSO)

Table 3.1 Classification of volcanic eruptions and associated products

Type of eruption	Type of Magma	Effusive activity	Explosive?	Product
Icelandic	Basic (low η)	Thick, extensive flows from fissures	Weak	Lava fields/plains
Hawaiian	Basic (low η)	Extensive flows from vents	Weak	Broad shield/dome
Strombolian	Basic + acidic (mod η)	Flows + ash	Weak to violent	Cinder cones and lava flows
Volcanian	Acidic (high η)	Flows absent, but if present thick	Moderate	Ash cones and explosive craters
Vesuvian	Acidic (high η)	Flows absent, but if present thick	Moderate to violent	Ash cones and explosive craters
Plinian	Acidic (high η)	Flows absent, but if present variable thickness	Very violent	Pumice, ash, lapilli, generally no cone
Peléan	Acidic (high η)	Flows absent, but if present variable thickness	Very violent + nuées ardents	Domes, cones of ash and pumice
Krakatauan	Acidic (high η)	Absent	Cataclysmic	Large explosive caldera
Rhyolitic flood	Acidic (high η)[a]	Small	Involves widespread ash flows over large areas	Flat plain or broad shield often within a caldera

Based on information in MacDonald, G. A., (1972) Volcanoes, Prentice-Hall
η = viscosity
[a] Can be fluid when not degassed

have been more explosive eruptions, but the scale of the caldera and the huge loss of life makes it stand out. The eruption was heard 3110 kilometres away in Perth, Western Australia, and at Rodrigues near Mauritius, 4800 kilometres away. At least 36,417 people lost their lives both as a result of the eruption and the tsunamis that it triggered. The impact extended much further as the sulphur and ash encircled the Earth in the upper atmosphere causing vivid sunsets and a short-lived deterioration in climate. It has been suggested that Krakatoa eruption was the inspiration for Edvard Munch's The Scream. The reddish sky in the background is the artist's memory of the deeply tinted sunsets which resulted for months during 1883 and 1884, about a decade before Munch painted The Scream (Olson et al., 2004).

It is worth noting that if the magma volume is much less and ground water is involved then the explosion from acidic volcanoes can be considerable without creating a large volcanic cone. In these cases, a maar is produced, essentially a deep, narrow crater with either a small rampart of ash or none at all. These craters are now usually lake filled, such as those of the Eifel Volcanic Region of Europe.

- Task 3.2: Find out about the Eifel Volcanic Region, when where the volcanoes last active, what are the typical dimensions of some of the maar craters?

Alternatively, you might like to find out about the recent Icelandic eruption Fagradalsfjall. This eruption started in March 2021 and there is some excellent citizen science reporting on YouTube

Further Reading[1]

MacDonald, G. A. (1972). *Volcanoes. A discussion of volcanoes, volcanic products, and volcanic phenomena*. xii + 510 pp., 120 figs, 144 pls. 15 tables. Prentice-Hall, International, New Jersey.

Olson, D. W., Doescher, R. L., & Olson, M. S. (2004). When the sky ran red: The story behind the "Scream". *Sky & Telescope, 107*, 28–35.

Smithson, P., Addison, K., & Atkinson, K. (2013). *Fundamentals of the physical environment* (4th ed.). Routledge.

Winchester, S. (2004). *Krakatoa: The day the world exploded*. Penguin UK.

---

[1] Almost every physical geography textbook contains a section on volcanoes and Smithson et al. (2013) *Fundamentals of the Physical Environment* is no exception. There is a popular account of the Krakatoa eruption written by Simon Winchester which is a good read.

Chapter 4
Earth's External Heat Engine

Source, Receptor, and Filter

Life on Earth is possible because of the Sun. We get light and heat from the Sun, but it is the geographical distribution of both around the Earth that makes our planet dynamic and subject to change.

This simple system involves a heat source (the Sun), a receptor (the Earth) and a filter (the Earth's atmosphere). We can ignore the vacuum of space, although not the distance between the source and receptor. Let us look at the impact of distance first. If we travel to the Sun's surface, and somehow avoid getting fried to a crisp, the surface temperature would be around 6000°K. Let us set up quadrat of known area, a small square if you like, and measure the heat (radiation) emitted over the area of the quadrat. Our measurement would have units expressed as radiation per unit area. Now this radiation (heat) spreads outwards from the source because the Sun's surface is curved. If this is challenging, then take a balloon and put two dots on it before blowing it up. As the balloon inflates the dots move further apart. If we keep the area the same as we move away from the sun the amount of radiation captured in the same unit area will fall. So, with distance from the sun the amount of radiation in the same unit area falls and so effectively it gets colder. Think of a child's picture of the Sun with rays radiating outwards, the distance between each ray is close at the surface but is greater as you move away from the surface, they spread out. In the vacuum of space, we don't lose radiation just spread it out over an ever-greater area as we move further from the Sun. That is why planets such as Mercury close to the Sun receive more radiation in a unit area than the Earth, or Pluto in the outer reaches of our solar system, does. We talk about the solar constant as the unit of heat received from the Sun just outside our atmosphere and it is approximately 1367 W/m² and varies by only as 0.25% during an 11-year Solar Cycle of sunspot activity. It is important to note however that in geological terms this is not a constant at all but has been slowly increasing since the Sun was formed and will continue to increase in the future.

© The Author(s), under exclusive license to Springer Nature Switzerland AG 2022
M. R. Bennett, *Our Dynamic Earth: A Primer*,
https://doi.org/10.1007/978-3-030-90351-0_4

The Earth is on average 150 million km from the Sun, with light taking 8.3 min to reach us, and it occupies the so-called 'Goldilocks' zone around the Sun, not to hot, not too cold, in fact just right for carbon-based life. If our Sun, which is quite an average star in the Universe, and a middle aged one at that, was hotter or colder the Goldilocks Zone would encompass a different selection of planets (i.e., be either closer or further away from the Sun). The Goldilocks Zone is also known as the Habitable Zone although this depends on what the requirements of life are. These may be different from those required by our planetary ecosystem. Finding the Habitable Zone around other stars is the first step in the quest for exoplanets and alien life.

So back to our simple system, source, distance, receptor, and filter, let us now think about the receptor. If the Earth's surface is shiny then the radiation received from the Sun will be reflected back to space. An ice sheet, or snow-covered landscape, would do a good job since it has a high reflectivity, like a mirror. Alternative dark surfaces absorb radiation, so a forest tends to absorb more radiation than a bright desert. Clouds also impact on reflectivity; more clouds, more incoming radiation is reflected back to space before it reaches the ground. We measure surface reflectance via something called albedo, the higher the albedo the more reflective the surface is. Land and water also have different thermal inertia. Rocks warm quickly, while not great conductors they are better than water. So, rocks (i.e., land) warm quickly but also cool quickly, that is they have little thermal inertia. In contrast water is less thermally conductive because the molecules are more widely spaced, and so water heats and cools more slowly. That is why it is always better to go swimming in the sea at the end of the summer when the water has had time to warm up.

There is one other big item to consider here and that is the Earth is not flat! Take an electric torch, or a flashlight if you prefer the American term. Turn it on and shine the beam vertically down on your desk, or on the floor. The beam of light is circular, if you now tilt the torch so the beam hits the surface at an angle the area of light made by the beam grows. The same unit of light is now spread over a greater area. If we think of the Sun now as the flashlight and its beam hits the Earth's surface overhead at the Equator it will be circular and focused, but towards the poles it will hit at an angle due to the curvature of the Earth and the area cast by the beam will be greater. So, at the poles the beam, or a unit of Sun's radiation, is spread over a larger area than when it is overhead at the Equator. This is why it is warmer at the equator than at the poles. Same amount of heat in the beam but spread over different unit areas.

Just like flea on the back of an elephant cannot tell if the elephant is doing a pirouette or not, it is hard for us to know that our planet spins with a tilt, but it does. Earth has an axial tilt of 23.5° at the moment and the geographic North Pole points away from the Sun. This gives us 24 h of daylight in the summer months in the high arctic and conversely 24 h of night during the winter. The reverse is true in the Antarctic. This difference in the length of the day impacts on the distribution of heat across the Earth's surface. The Sun is overhead at the Equator only during the two equinoxes, one in September and the other March. They are equidistant from two solstices when the Sun is over head at the Tropic of Cancer and Capricorn, respectively.

As the heat radiation reaches the Earth's surface it will warm it. As it warms it will in turn radiate heat back towards the atmosphere. The average temperature of the Earth's surface is 300°K. Now according to something called Wein's Law the warmer a surface is, the shorter the radiation emitted. So, the Sun is hot and therefore the radiation we receive will have a short wavelength, while the Earth is cool and consequently emits long wave radiation. If you are into laws of physics, you could also note one by Stefan–Boltzmann at this point which basically says that hot things give out more radiation than cooler things. Some laws of physics simply dress stuff up that is blindingly obvious to anyone who has ever burnt themselves.

We have looked at the source, distance, and receptor, so our final element must be the filter. Our atmosphere plays a big role in letting stuff in and out. So, the short-wave radiation from the Sun cuts through the atmosphere like a sharp knife. Scattering of radiation by gaseous molecules (e.g., O_2, O_3, H_2O and CO_2) occurs and roughly half of this radiation is scattered and lost to space, the remainder is directed towards the Earth's surface from all directions as diffuse radiation. Oxygen and Ozone molecules absorb key short wavelengths and keep harmful UV light in check.

- Task 4.1: The presence of ozone is a critical part of life on Earth. What is the Ozone Hole and how has environmental activism helped safeguard it? Write some notes about what you find out

The long wave radiation emitted by the cooler Earth is affected by different gaseous molecules as it passes through the atmosphere and here methane, water vapour and carbon dioxide come into play as so-called Greenhouse Gases. These gases absorb longwave radiation and consequently warm-up and re-radiate heat back to the Earth. The more Greenhouse Gas in the Earth's atmosphere the more effective the atmosphere is in retaining heat; the better it is as an insulating blanket. This is the key piece science in causing Global Warming today; more carbon dioxide from the burning of fossil fuels, warmer the atmosphere becomes and in turn the warmer the Earth below must get. The importance of carbon dioxide on climate is such that some argue it is the key variable in understanding climate change throughout Earth history, not just in the present day, more on this a bit later.

- Task 4.2: What are the main sources of methane on Earth? Note that methane is four times as effective as a Greenhouse Gas compared to carbon dioxide. Why are people less concerned about methane than carbon dioxide?

The net result of all of this is that the radiation received at any point on the Earth's surface will vary by month and by latitude. This is not quite the same as the effective temperature because other factor plays into this like cooling winds and ocean waters, but it is a key climate variable. It is predictable and can be calculated for any point on the surface. Figure 4.1 shows a diagram of this, a contour map showing the highs and lows of received radiation by latitude and by month. This explains why it is hot at the equator and cool at the poles and it is a fundamental diagram in Physical Geography and a product of the simple system we have

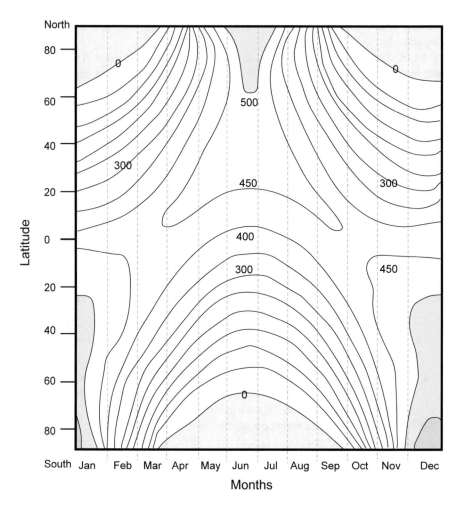

Fig. 4.1 Radiation (W/m²) received by the Earth plotted as contours by latitude and by month. This fundamental distribution of heat across the surface is key to understanding the seasons, zonal climate and the difference which drive atmospheric and ocean circulation. (Modified from: Neiburger, M., Edinger, J.G., Bonner, W.D., 1982. *Understanding Our Atmospheric Environment.* Second Edition, WH Freeman, Fig. 3.8)

described above. It is worth noting that if we were to change any of the variables in our simple model, we would change this graph. Since many of those variables, such as the surface properties of the planet or the atmospheric composition can all be easily changed, we should not be surprised that the Earth's climate is fundamentally quite unstable over time.

We can cut and dice this diagram in a slightly different way to create a second fundamental diagram. Let us average the radiation received over a year and plot this by latitude, let us also plot the radiation that is lost from the Earth's surface. We can see this as a debit and credit ledger by latitude (Fig. 4.2). At the equator, the ledger is in credit, more radiation is received than lost, but at the poles the reverse is true more is lost than gained and the ledger is in deficit. This is the core of the Earth's external heat engine and it is this imbalance in radiation that drives both the Earth's atmospheric and ocean circulation. It is also important to note that changing this ledger has an impact on the vigour with which both atmospheric and ocean circulation runs at.

Earth's Heat Engine

Let us look at the Earth's heat engine (Fig. 4.2) in a bit more detail, but we first need to start with a fundamental concept of our Universe, that of entropy. Entropy is the property of energy that involves its progressive decay into disorder; while I would not encourage such a definition in an examination basically it states, 'that everything goes to shit over time!' If you have a surplus of energy at one location, then it will decay over time to give a random distribution. Whenever you get a differential or potential difference then the system naturally moves to remove that imbalance. So, it is with the Earth's heat engine, we have a heat surplus at the Equator, a deficit at the Poles and so 'the system' natural works to remove that imbalance by transferring heat. A transfer is a flux (an amount of energy in a unit of time) and this naturally accumulates. Look at Fig. 4.2 and let us divide the surplus vertically into a series of columns say 2° of latitude each. We move the heat in the first column

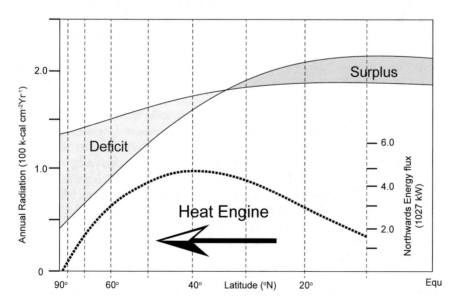

Fig. 4.2 Earth's heat engine. (Modified from Barry R.G. and Chorley R.J. (1982) Atmosphere, Weather and Climate, Fourth Edition, Methuen, London, Fig. 1.25)

northwards, and it adds to that in the second column, we move this double-load northwards, in the third column we have three loads and so on. It accumulates so the flux, the energy we need move, also accumulates, and will reach peak as shown in Fig. 4.2 just beyond the end of the balance point between surplus and deficit. The greater the energy flux, the greater the work required to transfer that energy. This peak in energy flux or movement occurs in the mid-latitudes, exactly where weather systems are most active.

Now Fig. 4.2 only shows half this engine, that for the Northern Hemisphere. We could assume that the Southern Hemisphere is the same and in broad terms it is but the deficit, surplus and energy flux are not symmetrical between the two hemispheres, however. The deficit at the South Pole is greater and therefore the energy flux, and atmospheric vigour, is greater. This is because the South Pole has the continent of Antarctica located over it. A polar continent both warms and cools more quickly than ocean water and the presence of an ice sheet means that it has a high albedo. The Arctic is an ocean basin and while it has a high albedo due to the sea ice the temperature regime is not as extreme as that over Antarctica. Antarctica is also isolated from other continents and surrounded by a polar current that runs uninterrupted around the icy landmass. There is a reason why we talk about the Roaring Forties, and they are a direct consequence of the greater meridional (north-south) energy flux in the Southern Hemisphere.

Change the surplus or deficit in either hemisphere and we change the energy flux and vigour of the heat engine. Again, this should make us think about how unstable the Earth's climate system is. However, before we get ahead of ourselves, we need to consider how this heat energy is moved. There are three mechanisms by which we can move heat; we can move sensible heat which you can sense and feel by wind. The cold breeze cools, while the warm breeze warms. This operates at a range of scales from local winds, and weather systems to global atmospheric circulations.

We can also move heat by moving warm or cold waters and ocean circulation is the key here. Finally, we can move heat via the process of evaporation and condensation. As water evaporates into vapour energy (latent heat) is taken up from the surrounding environment to energise the water molecule to break free. When we condense water vapour energy is given out. If we evaporate and condense the same package of water at different locations, that is we move the vapour on the wind, then we transfer heat. Energy is absorbed at the location of evaporation and given out at the location where condensation occurs. So, the third mechanism involves rainfall and latent heat. Changing the surplus or deficit of energy in either hemisphere must therefore change the vigour of atmospheric and ocean circulation. Perhaps you begin to see why Fig. 4.2 predicts many of the Earth's most dynamic external systems, including climate change, yet I am sure you will agree that it looks quite dull!

To understand how atmospheric circulation works we need to understand the origins of wind. This stems from difference is air pressure and winds blow from areas of high to low pressure to remove the differential. Imagine you are sitting on the beach at Bournemouth on a hot summer's day. The land heats up quickly in the sunshine and the air in contact with the ground expands. The gaseous molecules become more energised by the heat and spread out. Density, mass in a unit volume, therefore, falls and so does the atmospheric pressure. This air begins to rise upwards. Now the air in contact with the sea is colder, the water does not warm up as quickly as the land due to its greater thermal inertia. So as air over the beach raises and cold air moves on shore to replace it; that is air moves from higher pressure over the water onto the land. The net result is a sea breeze. Keep partying on the beach into the night and the reverse will occur. The land will cool more quickly than the sea. The cold air on land will sink; increasing the air pressure, but the water will remain relatively unchanged and become effectively warmer. The result is a land breeze, air move from areas of high to low pressure and pressure is determined by temperature (Fig. 4.3).

This simple system can be used to work up a model for atmospheric circulation we just need to add in two complicating factors. The first of these complicating factors is friction. The boundary-layer (earth-atmosphere surface) has trees, mountains, and buildings on it which provides friction and thereby impedes air flow. In the upper atmosphere there is no boundary-layer friction, and we can predict that circulation will therefore be slightly different than on the ground.

The second complicating factor is that the Earth spins beneath the atmosphere. This is referred to as Coriolis Effect which a frictional or inertial force. This is really all about the frame reference from which one views something. If we have an aeroplane at the North Pole flying south, when viewed from space they fly south in a straight line. But when viewed from the ground, which is rotating beneath the plane, the flight path appears to be deflected to the right. Repeat with a plane flying north from the South Pole and the deflection is to the left. If we now replace the poles with an area of high pressure adjacent to one of lower pressure, then air (wind) should flow from the high to low pressure areas but will be deflected to the right in the northern hemisphere and to the left in the southern hemisphere. Wind rather than blowing from high to low pressure is seen therefore to move parallel to areas of high and low pressure. We can contour pressure, a line of equal pressure is called an isobar, and according to the above wind will move parallel to these isobars. That holds for the upper atmosphere but close to the ground we have friction, and the movement is therefore altered slightly, and the wind will blow in an oblique direction to the isobars.

Thinking globally, we have an excess of heat at the equator, which should correspond to low pressure, a deficit at the poles which should give high pressure and air should in theory move between the two areas thereby redistributing heat. In practice this is over simplified, and we end up with three main elements to primary or global atmospheric circulation. These are (Fig. 4.4): Hadley Cell, Upper Westerly's (Rossby Waves), and Mobile Polar Highs (Polar Cell). Some would argue that Monsoons need to be added to this list, and we will do so here since they are of global significance. Cartographers like to draw pictures of these circulations and depict them with static lines and arrows, but in truth all of these elements are dynamic and subject to variation over time.

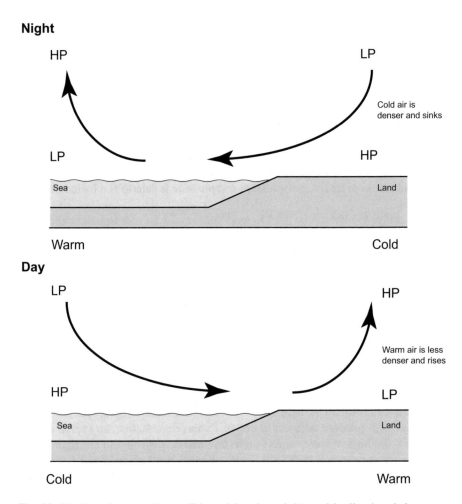

Fig. 4.3 Simple land- and sea-breeze. This model can be scaled to explain all major wind systems, that is wind blows from areas of high to low pressure

At the Equator, the Sun is overhead during both equinoxes and maximum radiation is received. The boundary layer will heat up and rise being warmer and therefore less dense than the surrounding air. As it raises the air cools and the water vapour within it condenses, towering clouds and tropical rainfall results. As the air reaches the Troposphere it suddenly finds itself colder than the overlying air of the Troposphere and moves north (or south) before descending over the sub-tropics. As it descends the air warms and give areas of intense high pressure, known as the subtropical high-pressure cells. In the Northern Hemisphere this occurs over the Sahel region of Africa. The descending air now moves back towards the equator (northwards/southwards depending on the hemisphere). As it moves it is deflected by Coriolis forces and forms the southeast or northeast trade winds. By the way, a

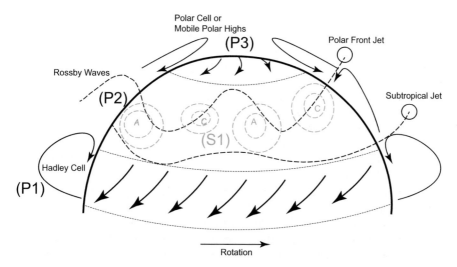

Fig. 4.4 Patterns of primary and secondary global circulation. P1 to P3 are elements of primary circulation and S1 is secondary circulation due to weather forming systems

wind direction is always specified as the direction from which it has come not the direction in which it is heading. When it approaches the equator, it completes the cell and rises again.

This surface convergence is called the Inter-Tropical Convergence Zone (ITCZ). The trade winds were once vital part of marine navigation, and they are the reason that most European maritime explorers ended up in the Caribbean and in Central America. Now as the overhead Sun shifts between the Tropics of Cancer and Capricorn the ITCZ also moves north and south with the seasons. The Hadley Cell is a powerful primary circulation and is effective in moving both sensible and latent heat in a meridional direction, but we need something else to continue the northward or southward heart flux.

In the upper atmosphere this is the westerly jet streams or Rossby Waves. This is a large movement of air from west to east that encircles the globe. The flow is parallel to the high-pressure area of the subtropics and rising air in the mid-latitudes. You might well wonder how effective this is in north/south heat transfer since it encirlces the globe. However, these westerlies in the upper atmosphere undergo a meandering cycle. Overtime the westerlies begin to meander and the amplitude of these meanders increase progressively until they collapse. The greater the amplitude the more effective is the meridional heat transfer. The meanders are partly controlled and stabilised by mountains such as the Rockies in North America and the Himalayas in Asia which help anchor the bends. In fact, the growth of both mountain ranges helped define the current Rossby pattern. On the surface the westerly's run in a more oblique direction across the mid-latitude oceans as westerly winds, although these surface winds are modified by secondary circulation patterns which we will discuss in moment.

The final element of primary circulation is mobile polar highs. Many textbook pictures of atmospheric circulation shows the southward movement of cold air as a series of arrows referred to in some places as polar easterlies or as the polar cell (Fig. 4.4). Cold air sinks over the poles and moves south or north to meet the westerlies. In practice this is less a continuous stream of air and more a series of polar high-pressure cells which break south from the Artic and north from the Antarctic. Imagine cold air sinking above the Arctic Circle and as it sinks being deflected, to the right in the Northern Hemisphere. It forms large high-pressure cells, with a clockwise air flow that break southwards periodically. These are the mobile polar highs and are more dynamic and periodic than the simple arrows drawn on many diagrams of atmospheric circulation.

This three-part system, Hadley Cell, Rossby Waves and Mobile Polar Highs completes our picture of primary circulation. This primary pattern of circulation also generates more local weather systems which also help move and mix heat around the globe. The best example of this is the succession of near surface low- and high-pressure systems that track eastwards across the Atlantic and Pacific oceans. Live in the UK and the daily weather you experience is defined by these systems, which involve the movement of both water vapour and sensible heat. Probably the simplest way of thinking about these is via an analogy with a spinning top or coin. Set it spinning, in either direction, and when it is released it will move along the surface until its momentum is exhausted. The same is true of weather forming systems in the mid-latitudes they are set spinning by convergence or divergence in the westerlies as we will see in a later chapter.

Some people consider monsoonal systems to be persistent global features and therefore part of the primary rather than the secondary (regional) pattern of circulation. Monsoons certainly occur over large areas of the globe and are persistent features of the current land-sea distribution. The summer monsoons produce more than 80% of the annual rainfall in some areas such as in India, Africa and Australia, and the percentage is more than 60% averaged across all global monsoon regions. They are a result just like land and sea breezes of differential heating over areas of land and sea. The Southeast Asian Monsoon is perhaps the best known, although the climate of much of Africa has a strong monsoonal component. The Asian Monsoon involves the rapid summer heating of the Indian sub-continent and surrounding areas. This drives uplift creating regional onshore winds which blow in from tropical seas. The combination of moisture laden air and rapid uplift drives intense rainfall – the monsoon. During the winter, the land cools rapidly, while the sea remains warm and consequently the regional wind flow is off-shore and on-shore rainfall is limited. Monsoonal systems reached a mega-scale during the Permian when all the continental area was concentrated into the supercontinent Pangaea. The land-sea difference in temperature drove intense seasonal rainfall and storm events.

- Task 4.3: Find out about the African Monsoon, how does it control rainfall over the year in Africa? Find a couple of maps of the monsoon system to supplement your notes

So far, we have concentrated on sensible heat transfer via winds, but movement of heat by the action of latent heat is also in play in most of these systems. Monsoons are a good example of this as is the Hadley Cell. Heat transfer via latent heat works on basis that evaporation requires latent heat, that is heat energy from the surrounding environment. Evaporation causes local cooling as heat is consumed, equally condensation gives out heat and leads to warming. If evaporation and condensation occur in different locations, you have moved heat. In the case of the Southeast Asian Monsoon, you have evaporation over the Indian Ocean and precipitation on the northern plains of India.

The other component of heat transfer is via the movement of ocean water directed by ocean currents. Given that ocean water covers about 71% of the Earth's surface it is quite important in the re-distribution of heat. The trouble with surface currents, which are the consequence of wind, is that they only really move the top 100 m or so of the water surface. Given that the Earth's oceans are on average 3688 m deep that is rather small surface layer. Subsurface current as we will see shortly are driven not by wind but by salinity. Let us stick with the surface waters for the moment.

The internet is a wonderful thing, and you can find any number of satellite images, maps or animations that depict ocean currents, yet one of the most elegant I have found is the map in Fig. 4.5. This is an old cartographic map produced by the US Navy in 1943 and it shows the key ocean currents. As you can see these form large gyres (circulations) in each hemisphere which move heat north and south although perhaps not as efficiently as one might hope. So why do we get these giant gyres?

To answer this question, we must follow a series of logical steps, but first we need to understand that sea level is not always level! There are many reasons for this. For example, gravity varies slightly across the Earth's surface and with it the water surface, tides also cause variation in the elevation of water and so does wind. It is the wind that is key here. So, wind blows over the surface of the sea and frictional drag cause the movement of water in the same direction as the wind. The

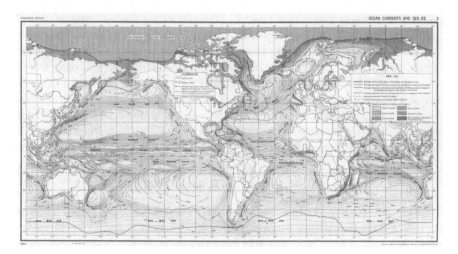

Fig. 4.5 Pattern of near-surface ocean currents. Red currents are warm and green currents are cold. (Public domain: United States Army Service Forces – Ocean Currents and Sea Ice from Atlas of World Maps. United States Army Service Forces, Army Specialized Training Division. Army Service Forces Manual M-101. 1943 from Perry-Castañeda Library Map Collection World Maps)

dominant near-surface winds in current generation are the trade winds and the westerlies. Now water dragged by these winds is deflected to the right in the northern hemisphere and to the left in the southern hemisphere by the Coriolis Effect. Focusing on the northern hemisphere we have the northeast trade winds and the westerlies both of which deflect water to the right. These two wind systems blow in opposite direction but at different latitudes. Crudely this will cause water to build up in elevation between these two wind systems around 30°N. The same occurs in the southern hemisphere. Now the crudest way to think of this is like a slight rise or hill of water. Water built up in this way will attempt to flatten this gradient by flowing from areas of high to low elevation, as it does so it in turn is deflected to the right in the northern hemisphere by Coriolis Effect. The net result is a huge gyre or circulation around the raised water level. These gyres dominate near-surface circulation but are modified by the shape of the ocean basins and as such they look slightly different in each ocean (Fig. 4.5).

These surface currents were once vital to shipping and can still save fuel on motorised vessels, but they are relatively shallow in terms of planetary air conditioning. There is one exception to this and that is their role in ocean turn-over. Look at Fig. 4.5 and focus on the western seaboard of South America. The currents are green (cold) moving from the Antarctic, but also run first parallel to and then away from the coastline. As they move away from the coastline, they take surface waters with them and this forces cold deep water to rise to effectively fill the space left. These upwellings are important in causing the oceans to over-turn, without them the oceans would tend to stratify and all, but surface waters would quickly become anoxic. They also provide huge quantities of nutrients to shallow waters and to the associated food web (plankton>fish>sea birds).

As someone once said there is 'money in muck' or in this case shit, guano if you prefer. All those sea birds of the coast of South America built up huge deposits of guano. The same is true of the coast of Namibia where there is similar coastal upwelling. The Guano Age made people rich and played a pivotal role in the development of modern input-intensive farming. Demand declined in the late nineteenth century with the discovery of the Haber–Bosch process of nitrogen fixing which paved the way for synthetic fertilizers. One of the initial foci where the Chincha Islands a group of three small islands 21 kilometres off the southwest coast of Peru. Whaling vessels carried consumer goods to Peru and returned with guano. In the peak year of 1870 Peru exported more than 700,000 metric tons of guano, although the resource was beginning to run out.

- Task 4.4: What can you find out about the Namibian off-shore guano deposits? What are the ocean currents that are key to this nutrient hotspot?

As an aside the export of guano from Peru to Europe may have played a role in the Irish Potato Famine and some have argued that the virulent strain of potato blight that caused such hardship may have originated in the Andean highlands. During the famine (1845–1849), around one million people died and a million more emigrated, causing Ireland's population to fall by between 20% and 25%. The Guano Age ended with the War of the Pacific (1879–1884), which saw Chilean marines invade coastal Bolivia to claim its guano resources. Knowing that Bolivia and Peru had a mutual defence agreement, Chile also mounted a pre-emptive strike on Peru, resulting in its occupation of the Tarapacá, which included Peru's guano islands. With the Treaty of Ancón of 1884, the War of the Pacific ended. Bolivia ceded its entire coastline to Chile, which also gained half of Peru's guano income from the 1880s and its guano islands. The conflict ended with Chilean control over the most valuable nitrogen resources in the world and as a result Chile's national treasury grew by 900% between 1879 and 1902. All because of a natural coastal upwelling!

Wind-driven ocean currents only impact on near-surface waters, those within 100 m or so of the surface. Deep water ocean currents are controlled by something else. Seawater is salty but that salt concentration can vary depending on the balance between evaporation which concentrates the salt, and freshwater input which dilutes it. Sea ice, formed when the ocean surface freezes, also adds to salinity because the process of freezing excludes brine. The salinity of ocean water varies with plate tectonics. The Gulf of Mexico is located in the sub-tropics which favours evaporation and salinity, moreover the major fluvial input from the Mississippi River is also rich in solutes derived from its huge catchment. It is not surprising therefore that the waters of the Gulf of Mexico are particularly saline. Contrast this with the coast of Alaska where temperatures don't favour evaporation and there is a large freshwater input from the coastal mountains that receive lots of rainfall. The primary controls here are all plate tectonic and over time these will change. The current pattern of deep-water circulation probably originated sometime in the Eocene and is known as the Thermohaline conveyor or circulation.

This conveyor is driven by the production of North Atlantic Deep Water (Fig. 4.6). Surface waters in the North Atlantic are saline, derived originally from the Gulf of Mexico and cooling northwards. The growth of seasonal sea-ice adds to this salinity. The cold, saline water is dense and therefore sinks and as it does so it pulls a current of warm surface waters up the Atlantic. This is one of the reasons that Northwest Europe is so mild and temperate in terms of climate. The deep water flows south down the Atlantic towards Antarctica. Deep water is also produced around Antarctica and together these cold bottom water moves northwards over the floor of the Indian and Pacific Oceans where they rise and return as surface flows to complete the circulation. When we talk about surface flows, we are talking about the surface minus the top 100 m or so where we get windblown currents. These will sometimes work with the underlying circulation such as in the North Atlantic and sometimes in opposition. The Thermohaline Conveyor involves vast volumes of ocean water helping to move heat and also helping to oxygenate ocean waters maintaining their biodiversity. But as we will see in a later chapter if we change the salinity in some way either slowly over time via plate tectonics or via some form of short-term instability we can cause this circulation to strength, weaken, or stop all together and in the geological past the circulation may have been very different all together.

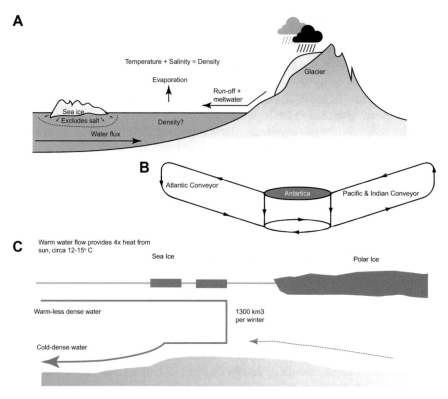

Fig. 4.6 Thermohaline circulation. (**a**) Controls on the salinity of ocean water. (**b**) Conceptual model of the Conveyor. (**c**) Impact of the Conveyor on the North Atlantic

Earth's Air Conditioning Systems

In the previous section we have seen how the imbalance of heat between the Equator and Poles drives the Earth's dynamic systems namely atmospheric circulation and near-surface ocean currents. Heat is also moved around the globe via sub-surface currents driven by salinity and temperature contrasts. It also follows that the vigour and health of these systems is determined by the temperature gradient between the Poles and Equator. If this temperature gradient was to change then we change the vigour of all of these Earth's systems. This ultimately is why we have climate change and why climate change is the normal state of our planet, stability is a much greater concern.

Further Reading[1]

Cockell, C., et al. (2008). *An introduction to the earth-life system*. Cambridge University Press.
Smithson, P., Addison, K., & Atkinson, K. (2013). *Fundamentals of the physical environment* (4th ed.). Routledge.

---

[1] Almost every physical geography textbook contains chapter that deal with atmospheric and ocean circulation. Smithson et al. (2013) *Fundamentals of the Physical Environment* provides a good summary of all of these basic concepts. Charles Cockell and colleagues gives a more advanced account in *An Introduction to the Earth-Life System* that I think is particularly good.

Chapter 5
Climate Change Is Normal

Climate and Earth History

You hear a lot in the media about climate change and how bad it is, and it is true humans will die and species will go extinct as a result and humans are to blame. However, what people forget is that climate *has always* changed throughout Earth history, often radically, and has caused species to go extinct in the past and this has radically changed the face of our planet.

If humans were not a significant source of climate change, as they are today, change would take place and species would go extinct. What is at issue is not the change itself but perhaps the *rate of change* and perhaps a human desire to make our planet remain the same as it is today or was in the recent past. That is impossible, nor desirable. Do not get me wrong, I am all in favour active advocacy for a more sustainable and equitable use of our planet's resources. But however hard you try you cannot, nor should you, prevent climate change since it is a natural process, but mitigating the role of human's in accelerating that change is just a matter of good stewardship for the short-time our planet is in our care.

It is all about time scales and one's perspective. Carbon dioxide has become a dirty word, a by-product of burning fossil fuels, yet life on Earth owes its very existence to this innocent gas. When the Earth was young, remember it is 4.6 billion years old, the Sun was much fainter and our planet too cold to sustain life. The Earth's atmosphere was full of carbon dioxide at the time and the greenhouse consequently much fiercer than today. This was life's salvation, it helped compensate for the faint Sun (Fig. 5.1). As I say it is all about perspective.

- Task 5.1: If you are interested in the idea of the faint Sun paradox then you could try looking at the original work by Sagan and Mullen (1972). For a modern review look at Feulner (2012)

© The Author(s), under exclusive license to Springer Nature Switzerland AG 2022   51
M. R. Bennett, *Our Dynamic Earth: A Primer*,
https://doi.org/10.1007/978-3-030-90351-0_5

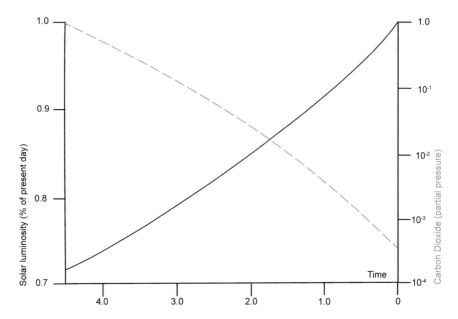

Fig. 5.1 The faint Sun Paradox. The solar constant has increased by 30% during the life history of the planet and carbon dioxide levels have fallen. Time is measured from the present day in billions of years. (Modified from Gretashum https://commons.wikimedia.org/w/index. php?curid=79674471)

We can explore climate change by thinking of three nested and therefore overlapping time scales. The first is measured in tens to hundreds of millions of years, stretching back from the early days of our planet to the present. Let us refer to this time scale as Megaanuum Plus, megaannuum meaning a million years. On this time scale the Earth's climate has oscillated at various times between what we can describe as 'Icehouse' conditions, defined by the presence of polar ice, and 'Greenhouse' conditions where polar ice was absent and the average temperature warmer than it is today (Fig. 5.2). Some workers have looked for cyclicity in these oscillations suggesting for example they occur every 150 million years or so, but in truth this is open to debate. What is clear is that throughout the Phanerozoic at least climate has moved between Icehouse and Greenhouse states.

If we now focus in on the last one million years or so, and call this sub-megaaannum, we see a different pattern of climate oscillations. The current geological period, the Quaternary, started about 2.8 million years ago and during this time the Earth's climate has oscillated between glacial and interglacial periods. In the last million years of so glacial periods have involved the growth of large continental ice sheets across the northern mid-latitudes. These icy periods were separated by warmer conditions, interglacials when ice was confined to polar regions. The last glacial, referred to in the UK as the Devensian, terminated around 13,000 years ago when we entered the current interglacial known as the Holocene. This medium-length time scale is nested with the longer geological timescale; it has occurred

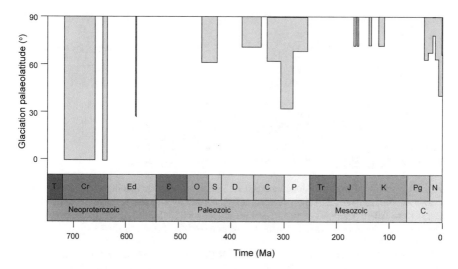

Fig. 5.2 Glaciations through geological time, or to be precise the last billion years. Intervals between these icehouse periods correspond to greenhouse states. (Mills, B.J., et al., 2017. Elevated CO_2 degassing rates prevented the return of Snowball Earth during the Phanerozoic. *Nature Communications* 8, 1–7. Fig. 1)

during the most recent part of the Cenozoic Icehouse. Other periods in Earth history show similar variations in global climate.

That takes us to the shortest of the three timescales, events that occur on a millennial scale. For example, the Middle Ages were associated in Europe with something known as the Little Ice Age, a period of time with severe winters that caused glaciers to advance in the European Alps (Fig. 5.3). At this end of the spectrum, we begin to merge into metrological trends and cycles. Climate is the long term, established weather patterns for an area, whereas metrological trends are variations in day-to-day weather. Trends over a few decades are really variation in metrology and only if they are sustained do, they become change to climate. The distinction here is a bit loose, but the point is one warm summer, or even several warm summers, does not signify climate change. That warming would need to be sustained over decades/centuries to become evidence of climate change.

The three timescales of climate variation are nested into one another and are also associated with different mechanism, but because they are nested those mechanisms are not independent of each other. For example, the millennial scale may be caused by subtle change in ocean circulation, the conditions for which are set by the medium scale factors, which are in term influenced by the mechanisms that determine climate on geological timescales. One potential unifying linkage is carbon dioxide, however.

- Task 5.2: What can you find out about the Little Ice Age? Find out some specific examples. How was it manifest in Britain or in another European country?

Fig. 5.3 A series of paintings done by the artist Rudolf Reschreiter that depict the Vernagtferner glacier as a fearsome beast. Although celebrating the glaciologist Sebastian Finsterwalder the idea that glaciers where beasts was common during the Medieval Little Ice Age and there are records of the Bishop of Geneva for example having to exorcise the Mer de Glace in 1645
Public domain:
https://commons.wikimedia.org/wiki/File:RESCHREITER_1911_Vernagtferner_01.jpg
https://commons.wikimedia.org/wiki/File:RESCHREITER_1911_Vernagtferner_02.jpg
https://commons.wikimedia.org/wiki/File:RESCHREITER_1911_Vernagtferner_03.jpg
https://commons.wikimedia.org/wiki/File:RESCHREITER_1911_Vernagtferner_04.jpg
https://commons.wikimedia.org/wiki/File:RESCHREITER_1911_Vernagtferner_05.jpg
https://commons.wikimedia.org/wiki/File:RESCHREITER_1911_Vernagtferner_06.jpg
https://commons.wikimedia.org/wiki/File:RESCHREITER_1911_Vernagtferner_07.jpg

Evidence for Climate Change

Before we get to far into trying to understand the mechanisms of climate change, we need to review some of the sources of evidence for that change. Perhaps the simplest way to evidence climate change is to try and observe it. In theory this makes sense, but it is not without its problems. If we travel to Iceland for example, the position of the margins of the Vatnajökull Ice Cap have been monitored on and off since the 1950s in some cases even earlier. This data is now available as part of the World Glacier Monitoring Service database and a simple internet search will take you to this.

Most glaciers show a retreat of glacier snouts, at least in Iceland. But is this simply a function of a centennial trend or long-term climate change? You have to remember that glaciers were advancing across Europe just a few centuries ago, during the Little Ice Age (c. 1645–1715).

Take another example, between 1968 and 1985 a series of droughts hit the Sahel region from West Africa to Ethiopia. Famine made famous by Band Aid. By the mid-1980s when the rains finally hit, approximately 100,000 people had died due to food shortages and disease. At the time this was argued as evidence of but is now thought to have been simply due to decadal variations in rainfall caused by period fluctuations in global sea surface temperatures.

There are no simple answers to these questions, other than to say the evidence must show a sustained, multi-centennial trend for it really to count as evidence of sustained climate change. A run of bad hurricanes does not, as the media would have you believe, indicate that our planet is becoming more hazardous due to climate change. This is one of the reasons why looking for evidence today of climate change is fraught with problems and subject to debate. The other problem with contemporary monitoring of change is that by the time you have assembled a long enough record to be sure the trends are real; it may be too late to do anything about them. As a consequence, the best evidence for climate change is actually to be found in the geological record; in fact we have had so much change in the last million years or so that the past forms a perfect natural laboratory to try and understand climate change mechanisms. There are clues in the landscape of climate change all around you.

Take a walk in any of the mountainous region of Britain and you will find evidence in the form of moraines and erratics (far travelled boulders) of the last glaciers to exist in Britain's uplands. These glaciers and small ice caps date from the Younger Dryas, sometime referred to as the Loch Lomond Readvance in the UK. These glaciers disappeared around 10,000 years ago, but their extent and mass balance (see Chap. 8) can be reconstructed from their remains. They give a picture of the past climate (palaeoclimate) at the time and since they decayed in the face of global warming at rates of 1° every ten years or so they provide analogues for how glaciers might respond to future global warming. We can reconstruct older ice sheets in the same way, like those that covered parts of North America and Fennoscandinavia during the height of the last glacial cycle. These glaciers reached their maximum extent around 20,000 years ago in what is referred to as the Last Glacial Maximum or LGM.

If glaciers are not your thing, then you can go to a desert, such as the Kalahari Desert in Botswana and look at vegetated sand dunes. These fossilised sand dunes record wind patterns before the dunes became stabilised by vegetation. Rivers also contain a record of past climate as we will explore in a later chapter. Coastal regions contain submerged and raised beaches or rock platforms which tell us about changing sea levels. The principal cause of recent sea level change at least is the waxing and waning of large continental ice sheets which store up ocean water. Wherever you look the landscape holds clues of past climate, the job of the geographer is to decode these proxies and understand their climate message.

Perhaps the best place to look for past climate signals is in organic deposits, like those that accumulate in lakes, marshes, or other wetlands. At these types of location, a range of organic proxies may accumulate. Pollen is perhaps the oldest of these proxies in terms of study, but today there is an ever-increasing array of sophisticated alternatives from the remains of beetles to midges. In theory the pollen that falls to the bottom of a water body should reflect the vegetation in the local area, may be a little wider due to the action of the wind. Each layer of mud may contain pollen, the oldest being at the bottom, and will provide an impression of the vegetation at the time the layer accumulated and the associated climate in which that vegetation grew. As climate deteriorated towards a glacial period so the vegetation slowly becomes more arctic in its character, the reverse being true at the start of an interglacial. The northward migration of warmer vegetation will be controlled by geography (barriers) and the location of plant refugia (locations where warmer species survived the harsh glacial period). The trouble with pollen is that it is a regional synthesis of the vegetation and may not accurately reflect the true ecological composition, it can also be slow to respond smoothing out abrupt climate changes.

In the 1970s and early 1980s the study of Coleoptera (beetles) was initiated and began to give excellent results. The climatic range of certain beetles is quite specific in terms of temperature and in some cases rainfall. Map out that range for a living species, then find it in the past and you have a climate proxy. Assuming that the climatic range of the species has not changed, when you find it fossilised in a layer of mud then the climate at the time the fossil lived must have been similar to a point in the modern climate range of that species. Adult beetles are distinguished from other insects by the presence of hardened forewings called elytra that cover and protect the membranous hindwings. It is the elytra that are usually found since they preserve well.

Atkinson et al. (1987) used beetles from a range of UK deposits to construct a high-resolution temperature curve using a neat concept. The concept was that of mutual climate ranges and involves the overlap in properties of two or more beetles each with a different climate tolerance. Find two or more different beetle species in a mud layer and the climate at that point in space and time has to lie within the overlap between the climate tolerance of those two species (Fig. 5.4). In this way you can be much more precise about the palaeotemperature.

There are many other organic proxies that can be deployed including diatoms, ostracods, molluscs, plant fossils, foraminifera, and Chironomidae. The latter is one of the newest proxies. The humble midge, the scourge of our wet uplands, will only

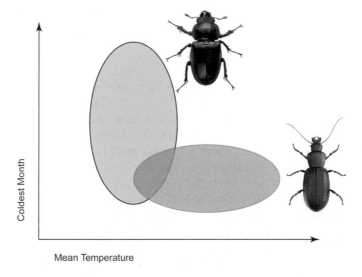

Fig. 5.4 Conceptual illustration of the idea of mutual climate ranges. Find two species of beetle (or any fossil for that matter) in the same layer the climate in theory must correspond to the overlap in climate tolerances of the two species

complete their life cycle if the environmental conditions are suitable. This life cycle is controlled by summer temperature, water acidity (pH) and the nutrients versus pollutants present. The heads of the larval stage are deposited in a marsh or lake floor after each moult as they grow. The heads build up year after year providing a natural archive of the Chironomid species that have lived in the water body. Researchers have determined the optimum summer temperature for over 200 common Chironomid species to create a 'midge thermometer'. Each layer of sediment represents a time in the past and the midge species that lived tell you the temperature to within ±1 °C.

Terrestrial records whether they be organic, or inorganic have one big problem, that is they are often subject to erosion. There are few places on land where sediment accumulates continuously for long enough to build up a long palaeoclimate record. Glacier's advance, rivers flood and coasts erode all causing hiatuses in deposition. If you want to find a continuous depositional record, then the best place to go is the deep oceans. A long way from land, out on the abyssal plain, terrestrial sediment does not reach these ocean floors. Here there is just the slow and steady fall of dead bodies from various planktons and other microorganisms. Little if anything breaks this cycle of sedimentation and a record of microscopic ocean life slowly accumulates over time and much of it has tiny hard parts made up of calcium carbonate that get preserved.

The 1940s saw the development of the piston core and in the 1950s custom built drill ships were developed able to stay on station above a sea floor target. These developments led in 1966 to the Deep-Sea Drilling Project which has continued in various forms since. The current iteration, International Ocean Discovery Program, commenced in 2013. This multi-decadal programme led initially by the US National Science Foundation has recovered literally thousands of kilometres of ocean bottom sediment cores. It was a team in Cambridge led by Nick Shackleton, a distant relative of Ernest Shackleton the famous explorer, who help made these ocean sediments central to our understanding of Quaternary climate change, however.

Urey (1947) first suggested that oxygen isotopes might fractionate with temperature. Throughout the 1960s recovered sediment cores where being analysed for the oxygen isotopes stored in the tiny calcium carbonate shells of foraminifera. The aim was to create a palaeothermometer. The uncertainty initially at least was the relative importance of temperature versus global ice volume in this fractionation. Nick Shackleton, and his US colleagues Jim Hays and John Imbrie published their now famous paper in *Science* in 1976 which showed that the ocean cores contained a record of global ice volume through much of the Quaternary. We will return to this seminal paper a little later, but first we need to understand how oxygen isotopes in fossil microorganisms on an ocean floor can tell us about global ice volume.

Oxygen has two stable isotopes oxygen-16 and oxygen-18 the only difference is the presence of two additional neutrons in oxygen-18 making it slightly heavier. On the ocean surface evaporation of water (H_2O) favours the lighter oxygen-16 and as a result water vapour is slightly enriched in oxygen-16. It falls as rain and returns to the ocean. The ratio of $^{18}O/^{16}O$ is maintained, we measure this ratio via $\delta^{18}O$, where δ is delta meaning rate of change. The oxygen in the calcium carbonate of a sea creature living in the ocean at this time will have a normal $^{18}O/^{16}O$ ratio. Now if a large ice sheet grows then some of the water vapour will fall as snow and be stored in the ice sheet rather than be returned to the oceans. As a result, the ratio of $^{18}O/^{16}O$ in the oceans will change, reflected off course in the sea creatures building carbonate shells. So, by looking at the layers of dead sea creatures on the ocean floor and extracting their oxygen isotope ratio we build up a $\delta^{18}O$ record (i.e., change in oxygen-18) and have a record of changing global ice volume. We can date the cores but also work out their age from the rate at which sediments accumulate at the site in question. This simple proxy has transformed our understanding of climate change. The synthesis of lots of marine isotope records (Fig. 5.5) allows us to build up an extended picture of global ice volume and therefore climate change throughout out the Quaternary Ice Age.

The quality of this record is such that we now recognise Marine Isotope Stage. Working backwards from the present, which is MIS 1 in the scale, stages with even numbers have high levels of oxygen-18 and represent cold glacial periods, while the odd-numbered stages are troughs in the oxygen-18 record and represent warm interglacial intervals. When you first look at this record (Fig. 5.5) what is striking is that there are lots of glacial and interglacial periods, far more than we have a terrestrial record for. This is because ice sheets and other source of erosion remove earlier parts of the terrestrial record, so it is incomplete.

The marine isotope revolution was followed quickly by another. In the 1980s the technology was advanced to drill cores through ice sheets. Solid ice cores extracted in this way contain a record of past climates. On the top of an ice sheet snow falls each winter, and then melts in the summer before refreezing. This creates two layers each year and gives what we call a couplet. Winter snow fall creates a white layer, due to the higher volume of air, whereas summer melting tends to form clear ice since the air is lost by melting and refreezing. By counting the couplets, you have a way of dating the layers. Now if you can count back to a particular year and in controlled conditions melt the ice from that layer you have a huge amount of information.

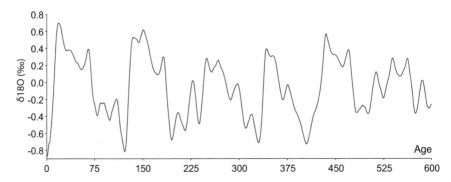

Fig. 5.5 Section of oxygen isotope record. A positive delta ^{18}O indicated less ice while a negative value indicates more global ice volume. (Plotted from SPECMAP data available from: https:// www.ncdc.noaa.gov linked to Imbrie J. et al., 1993. On the structure and origin of major glaciation cycles 2. The 100,000 year cycle. *Palaeoceanography* 8, 699–735)

Firstly, the gas you release when you melt a layer comes from tiny air bubbles in the ice and represents a perfectly preserved sample of the atmosphere at that time. This allows you to measure methane and carbon dioxide levels directly. You can also test the electrical properties of the ice layer which tells you about atmospheric dust at that time; the higher the conductivity the more dust there was. Volcanic eruptions stand out very clearly in this way mainly due to the sulphur dioxide the deposit in the ice. However, the real power comes from the oxygen in the water; the oxygen isotopes tell you about temperature. Unlike the oxygen isotope in marine cores those from ice cores can be calibrated to give a direct indicator of temperature at the time of evaporation.

Now you might be wondering about what happens to these layers of ice when the ice flows, as glaciers and ice caps do. Well, the simple answer is they get deformed, and the record is lost, but if you drill your hole down from the highest point on an ice cap the point from which ice flows outwards in all direction then you can extract an undisturbed core.

In the late1980s two ice cores were drilled 30 km apart from the summit of the Greenland Ice Cap, one by an European consortium and another by an American team. The Greenland Ice Core Project (GRIP) sponsored by the European Science Foundation ran from 1989 to 1995 and successfully drilled to a depth of 3029 m. GISP-2 was drilled by a group of US Universities and reached bedrock at a depth of 3053 m. Both cores gave a climate record of about 100,000 years before they became to deformed. Perhaps the most remarkable story that came from these ice cores, both of which helped to further transform our understanding of recent past climates, was the scale and rapidity of climate change during the last 100,000 years. We will look at this further in a while.

- Task 5.3: It possible to use historical records to reconstruct climate before instrumental records were kept routinely. One particular source is ship logbooks. There are several papers which use this type of information, why not check out this paper by Barriopedro et al. (2014). Alternatively look at this paper by Wheeler et al. (2009) who reconstruct the path of a hurricane from 1680

In summary, there are lots of ways in which we can demonstrate how climate has changed at various scales and we have touched on just a few of the available methods and proxies.

Mechanisms of Climate Change

Let us now introduce a simple model of climate change (Fig. 5.6). This model recognises four main agents. The first is variation in the amount of solar radiation received. We have already seen how the sun has become brighter over geological timescales, and how the seasons are caused by variation in the receipt of solar radiation due to the Earth's orbital parameters. These parameters can vary over time.

Fig. 5.6 Simple model of
climate change that
recognises four variables,
labelled one to four: land
surface properties (1);
oceans (2); solar input (3);
and atmospheric
chemistry (4)

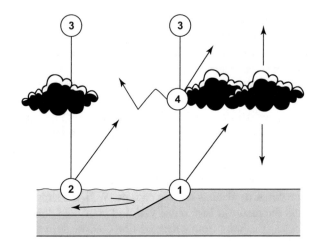

The next set of agents are based on the land surface itself. These may vary with
changes in landcover. Deserts and grasslands, for example, have different albedos,
as does snow covered land. We also have a whole set of plate tectonic factors that
operate to change the distribution of land and sea, as well as the shape of the ocean
basins. Finally, there is the atmosphere itself and its chemical composition. This
influences effectiveness of the greenhouse. We will explore each of these in turn
before looking at some planetary scale feedbacks that might accentuate some of
these causes.

Landscape and Land-Sea Distribution (1)

This group of factors include ones that operate on relatively short-time scales such
as changes in land cover as well as ones that operate on geological timescales and
involve plate tectonics.

Landcover can have a profound impact on surface reflectivity (albedo) and it is
influenced by the distribution of deserts, grasslands, ice sheets and tropical rainforests. Basically, by landcover. Change this and you change the local heat balance.
Trees, or more precisely tropical rainforests, have other impacts too. Forests affect
the climate in three ways: by absorbing carbon dioxide and thereby reducing greenhouse gas; by evaporating water that forms clouds which increase atmospheric
albedo; and by absorbing sunlight with their dark leaves, which warms the Earth. In
high latitudes the effect of the dark leaves out weights the cooling benefits, but in
tropical rainforests the biological absorption of carbon dioxide is much greater due
to the greater biomass involved and evaporation is more intense. This leads to more
clouds which reflect sunlight back to space. Tropical rainforests act as giant planetary air conditioners.

Many landcover impacts can be quite complex. Take bushfires for example they release carbon dioxide, but also lead to cooling because soot within the atmosphere helps scatter radiation. The latter is part of a general idea referred to as Global Dimming and is seen by some to work in opposition to Global Warming. Sulphates derived from burning of fossil fuels, bush fires, and volcanic eruptions all contribute condensation nuclei and therefore aid condensation and cloud formation. Clouds increase atmospheric albedo and the reflection of incoming radiation. In theory the effects of global dimming have been declining since the late 1960 when the UK and other developed countries started to implement clean air acts. The irony is that pollution can sometime help slow climate change.

Perhaps the most elegant illustration of the power of albedo to control planetary temperature was developed by James Lovelock and his colleagues in support of his idea of Gaia. The Gaia (Greek Deity, Mother of all Life) concept suggests that the Earth should be considered as a living because it is able, like many organises, to regulate its own temperature at least to some degree. Lovelock developed the idea of Daisy World to illustrate this idea (Fig. 5.7). No one is suggesting that the humble daisy can actually control planetary temperatures as one of my students once wrote! It is simply a thought experiment to show how planetary regulation might work. The experiment works from the premise of a progressive increase in the solar constant through time. Two types of daisies are the only plants present. Initially the simulated planet is rather cold for life and black daisies thrive because they absorb more radiation. However, as the solar constant rises it becomes a bit too hot for the black daisies and white daises, which reflect more of the radiation, become dominant. In this way the gradual transition between the two daisy types controls the planetary

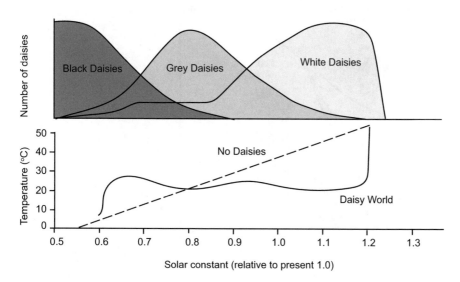

Fig. 5.7 The theoretical model/concept of daisy world used to explain how planetary scale feedbacks could work. (Lovelock, J.E. 1988., The Ages of Gaia –A Biography of Our Living Earth, W. W. Norton, New York. And Wood, A.J., et al., 2008, Daisyworld: A review, Review of Geophysics 46, RG1001, Fig. 4)

temperature. It is a *simulated* example of a planetary feedback loop, something we will come back to in a while.

- Task 5.4: Agent based models are a way of modelling natural systems. There is a daisy world simulation written for NetLogo which runs in a web browser try this link: https://rb.gy/rvspk4. The link should take you directly to a web version of the model, if not search for NetLogo and navigate to the online models. Click the grey window and the model should run. Play lay with the daisy variables to see how they impact on climate

In summary, the biosphere and its impact on the landscape can influence climate. However, most of the landscape variables operate on a much longer timescale, namely the time scales of plate tectonics. Figure 5.8 shows the latitudinal temperature gradient associated with various different continental configurations. Let us first contrast 'Polar Cap World' with 'Tropical Ring World', the former has much colder temperatures at the poles than the latter. The latter is also warmer at the tropics. This because land has less thermal inertia than ocean; it warms up quickly but also cools quickly. In addition, ocean water is darker and absorbs more heat. If we now add an ice sheet to the 'Polar Cap World' scenario we can see that the polar temperature falls even further. This is because of the high albedo associated with shiny ice and snow. Plate tectonics moves the continents around and therefore has a long-term impact.

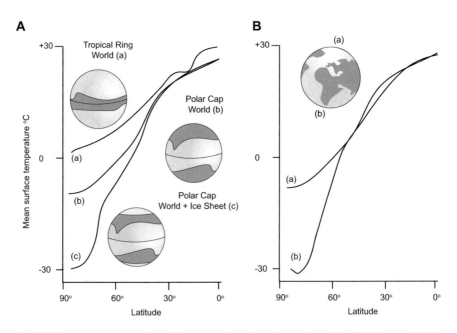

Fig. 5.8 Temperature curves for different continental configurations. (Modified from: Cockell, C. et al., 2007. An introduction to the Earth-Life System. Cambridge University Press, Fig. 3.13)

Plate tectonics also drives mountain building, which we will see in a later section influences the carbon dioxide content of the atmosphere. Mountains also deflect air flow. We saw in Chap. 4 how the Rossby Waves are a key part of global atmospheric circulation. Their position and their north-south amplitude is both influenced and partially stabilised by the Himalaya in Asia and the Rockies in North America. Using general circulation models, giant computer models of the Earth's atmosphere, the effect can be seen by modelling the circulation with and without surface topography. Both mountain ranges help stabilise the position and also increase the movement of warm wet air northwards. This is exactly the kind of air mass you need for snowfall in northern mid-latitudes and therefore for the growth of mid-latitude continental ice sheets.

Ocean Basins (2)

Ocean currents help to redistribute heat as we have seen before, and their geometry is influenced by the shape of the ocean basins. If you change the basin geometry via plate tectonics, then you change the currents and the efficiency with which heat is moved poleward. The conceptual configurations shown in Fig. 5.9 illustrate this point rather nicely.

We have discussed previously the importance of the thermohaline circulation in moving deep water around the World's oceans today. The circulation emerged in the late Cenozoic as a result of change in the ocean floor in the North Atlantic. Before this the circulation was reversed with downwelling being a feature of the northern Pacific rather than upwelling as today. The salinity balance of our oceans is influenced by the distribution of mountains around ocean basins, these control rainfall and runoff and consequently the freshwater input to nearby oceans. Today the Gulf of Mexico has an important role, like a giant saucepan, in helping to create warm saline waters that flow northward in the North Atlantic and cool.

Now the thermohaline conveyor plays an important role in modulating climate in the area around the North Atlantic (Fig. 4.5). Slow or shut the conveyor off and

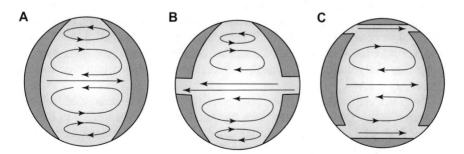

Fig. 5.9 Different theoretical continental configurations and their impact on ocean circulation and therefore heat transport between the Equator and the Poles. (Modified from: Cockell, C. et al., 2007. An introduction to the Earth-Life System. Cambridge University Press, Fig. 3.15)

Europe cools by the order of 10–15 °C such is the power of the warm waters. Now in theory one can modulate the strength of the conveyor by changing the flux of fresh water in the North Atlantic and also by changing the extent of sea ice. More freshwater and less sea ice and you have surface waters that are less dense and will not sink. If Atlantic Deepwater does not form, then the conveyor begins to break down and may adopt a different configuration. Some people believe that this is one of the key variables in understanding climate change during the Quaternary. We will return to the role of the conveyor as we try to build a picture of Cenozoic climate change later in this chapter.

Solar Input (3)

The orbital configuration of our planet is not as fixed as perhaps we think, or geography textbooks portray. Our orbit around the sun, the angle at which the Earth tilts and the direction in which it tilts is all subject to the gravitational pull of Saturn and Jupiter. The Earth moves within the solar system with a few regular wobbles as different gravitational fields align. This has two regular impacts, first it effects the total amount of solar radiation the Earth receives, but also secondly its latitudinal and seasonal distribution. Basically, it causes the patterns in Fig. 4.1 to change.

At the longest of timescales, the Earth's orbit around the sun changes from an ellipse to a more circular form. This happens on cycles of 400,000 and 100,000 years and the impact are a subtle change in the total amount of solar radiation received. With a cycle of approximately 40,000 the Earth's axial tilt varies from 22.1° to 24.5°, it currently has a tilt of 23.44°. This impacts on the latitudinal distribution of solar radiation received. Finally, there are several cycles around 22,000 years which relate to the direction in which axis tilts and this processes around a circle over time. One of the impacts is to vary the amount of solar radiation received in the tropics (Fig. 5.10).

These cycles where first accurately correctly calculated by Serbian geophysicist and astronomer Milutin Milanković (often Angelized to Milankovitch) in the early part of the twentieth century and advanced as a cause of the Ice Age. A Scottish scientist, amongst others, had postulated the idea in the nineteenth century, but struggled with both accurate calculations and a sceptical audience. Milanković also struggled to convince the scientific community that when combined these subtle cycles in solar radiation, and they are subtle, could have such a profound impact as to provide the pulse-beat of glacial to interglacial oscillation observed for the Quaternary Ice Age.

It was not until a seminal paper Hays et al. (1976) that the ideas of Milanković where validated. Deep sea cores from the Indian Ocean were subject to oxygen isotope analysis. Put crudely minute calcium carbonate shell fragments from different layers in core where crushed and the ratio of oxygen-16 to oxygen-18

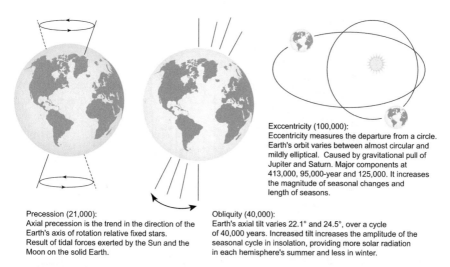

Exccentricity (100,000):
Eccentricity measures the departure from a circle.
Earth's orbit varies between almost circular and
mildly elliptical. Caused by gravitational pull of
Jupiter and Saturn. Major components at
413,000, 95,000-year and 125,000. It increases
the magnitude of seasonal changes and
length of seasons.

Precession (21,000):
Axial precession is the trend in the direction of the
Earth's axis of rotation relative fixed stars.
Result of tidal forces exerted by the Sun and the
Moon on the solid Earth.

Obliquity (40,000):
Earth's axial tilt varies 22.1° and 24.5°, over a cycle
of 40,000 years. Increased tilt increases the amplitude of the
seasonal cycle in insolation, providing more solar radiation
in each hemisphere's summer and less in winter.

Fig. 5.10 Three main Milankovitch cycles

determined. Changes in oxygen-18 relate to the volume of global ice at the time the shells lived in the oceans. The more ice the richer the oceans are in oxygen-18. Hay et al. (1976) subjected this time series of oxygen-18 to a spectral analysis which identified the common component cycles and found cycles at 100,000, 40,000 and 22,000 years that is they found the Milanković radiation cycles (Fig. 5.11). Since then, this type of study has been verified by numerous workers and with data from all the major ocean and accurate climate signals have been assembled for the last 65 million years.

Despite the fact that we now know that orbital radiation cycles drive the pulse beat of climate change at sub-megaannum timescales there remains a basic problem – the variations in radiation are small compared to the global scale of their impact. An amplifier is needed and a big one at that!

There are a few other puzzles. If we take a marine record for the whole of the Quaternary and break it down into sections, then the record seems to be modulated by different Milanković cycles at different times (Fig. 5.11). For example, during the last three quarters of a million years the dominant cycle is that of eccentricity. We see glacial periods which last for around 100,000 years punctuated by interglacial periods of 10,000 years. If we look at the earlier part of the record, then the glacial-interglacial oscillations seem to be driven by obliquity and alternate with a 40,000-year rhythm. Why have different cycles been more dominant in the climate record at different times? In addition, what about climate change that occurs at millennial time scales; at time scales below those of the Milankovitch cycles which have emerged from the ice core record.

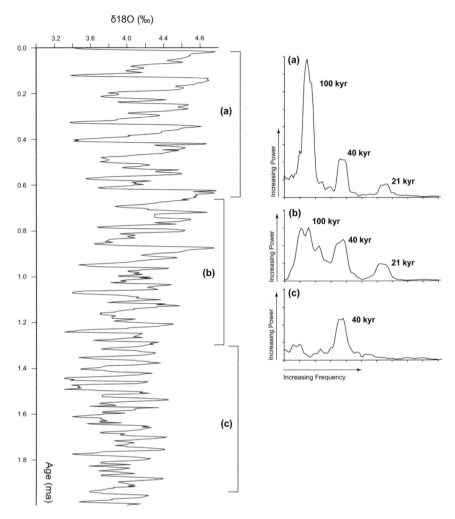

Fig. 5.11 Time series analysis on different sections of the Quaternary oxygen isotope record. Different Milanković appear to be dominant at different times. (Plotted from SPECMAP data available from: https://www.ncdc.noaa.gov linked to Imbrie, J., et al., 1993. On the structure and origin of major glaciation cycles 2. The 100,000 year cycle. *Palaeoceanography* 8, 699–735)

Atmospheric Content (4)

This mechanism of climate change resolves around the greenhouse effect and its dependence on carbon dioxide. Before we unpack this further it is worth saying that while the scientific press obsesses about carbon dioxide it is just one of several greenhouse gases. More potent is methane, four time more potent in fact, and water vapour also has a role to play. The reason that people are less concerned about methane is that it breaks down within a decade or so in the atmosphere, whereas carbon dioxide is more long lived. The principal sources of methane by the way are polar wetlands, tropical wetlands (paddy fields in particular) and farting livestock.

Whatever the greenhouse gas in question is, the greater its volume in the atmosphere the more longwave radiation, emitted by the Earth, is absorbed and consequently the warmer the atmosphere becomes. Some of this additional heat is radiated out into space, some is radiated back to Earth causing surface temperatures to rise. One point to remember is that the effect of a greenhouse is to warm the planet equally, not selectively. Just because one country may produce more carbon dioxide than another does not mean that the impact of global warming is felt differently, we all have a collective stake in our atmosphere. The point is important beside its relevance to the geopolitics of climate change, since in theory global warming should not accelerate the heat differential between the equator and the poles. As such it should not lead to an increase in storminess, as is often suggested, it might even reduce the meridional heat differential slightly. Given that the vigour of the Earth's oceans and atmospheric circulation depends on the size of this differential it is an important question. Not everyone agrees with this, and more research is needed but the case for an increase in extreme meteorological events due to global warming has currently not been made, a redistribution yes but a more deadly planet no.

Let us set aside the burning of fossil fuels for a moment and focus instead on the geological controls of carbon dioxide. The principal source of carbon dioxide in the atmosphere on geological time scales is volcanoes. While TV documentaries like to show broiling pits and explosions of red-hot lava the principal product of any volcano is gas. So, when volcanism increases so does carbon dioxide. As Pangaea began to break up in the Mesozoic lots of new mid-ocean ridges were formed which are effectively giant chains of volcanoes. The Mesozoic is one of the warmest periods in geological history as a consequence. Therefore, the breakup of supercontinents can lead to increases in carbon dioxide. Now plate tectonics also scrub (remove) carbon dioxide via mountain building. Chemical weathering of rocks removes carbon dioxide and the greater the surface area of rock the greater the weathering rate. Mountain building increases rock surface area. Rain falling through the atmosphere and in particular percolating through soil becomes weakly acidic:

$$CO_2 + H_2O \rightleftharpoons H_2CO_3$$

For a generic calcium silicate, we get:

$$2CO_2 + H_2O + CaSiO_3 \rightleftharpoons Ca^{2+} + 2HCO_{-3}^{-} + SiO_2$$

This works for almost any silicate mineral, not just for calcium carbonate (i.e., limestone and chalk). Weathering removes carbon dioxide and weathering is enhanced by warm, damp climates with lots of fresh un-weathered rock. In the Carboniferous, much of Britain and Europe were located in the tropics and vast tropical rainforests grew, although the trees were none flowering species. The tectonic setting at the time caused this carbon to be buried, forming the coal deposits that fuelled the industrial revolution and are now driving global warming. The burial of carbon in the Carboniferous was sufficient to drawdown atmospheric carbon dioxide levels such that the planet cooled and entered an icehouse state with a giant ice sheet over much of Gondwana (a continent made of Africa, South American, Australia and Antarctica). Today as we burn this fossil fuel, we are simply seeing today the flip-side of this ancient cooling.

In Fig. 5.2 we saw how planet Earth has oscillated between icehouse and green-house conditions throughout at least the Phanerozoic. It has been suggested that this reflects the dominant plate tectonic setup (Fig. 5.12). The contrast is between a scenario where most subduction occurs below island arcs, here the carbon dioxide output is modest. However, where subduction occurs mainly below continental margins carbon dioxide production is greater. This is because continental crust has limestone and chalk within it which when 'cooked' yield lots of carbon dioxide. The latter tends to occur when subcontinents are beginning to break up.

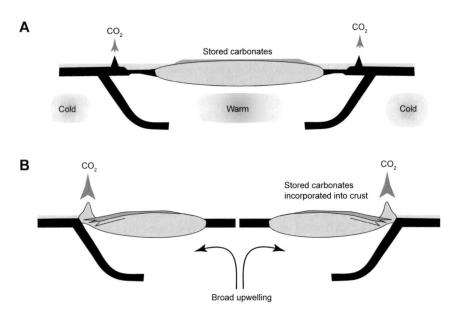

Fig. 5.12 Conceptual model of two different types of plate tectonic set up, the continental arc mode is associated with much greater production of carbon dioxide than the island arc mode and consequently tends to be associated with a greenhouse world such as existed during the Mesozoic. CO_2 emitted during the arc mode is derived from the mantle and from subducted carbonate. An island arc scenario might be favoured during continent aggregation that is the first half of a super continental cycle. During this time, the continental mantle below the arcs heats and the oceanic domain cools, setting the stage for a large mantle overturn event which breaks up the supercontinent. Mantle overturn triggers continental dispersal, driving the leading edge of continents to advance trench-ward, which places subduction zones into compression. Continental arcs and fore-land fold and thrust belts in the back arc region incorporate stored carbonates. As this carbonate is purged it amplifies the total volcanic emissions of CO_2 as well as total deposition of carbonate. (Lee C-T.A. et al., 2013. Continental arc–island arc fluctuations, growth of crustal carbonates, and long-term climate change. *Geosphere* 9, 21–36, Fig. 10)

Planetary-Scale Feedbacks

In the previous sections we have shown there are four main controls on climate that operate at different scales and more importantly are not independent of one another. The climate change history in the geological record is the product of all of these mechanisms working together.

There is another aspect to consider here which is that of non-linearity, which is an element of chaos theory. Edward Norton Lorenz (1917–2008) mathematician and meteorologist is most closely associated with the idea and the analogue most frequently used to both explain and mock the idea of chaos. There are various forms of this analogue from the famous talk title 'Does the flap of a butterfly's wings in Brazil set off a tornado in Texas?' at the American Association for the Advancement of Science in 1972 to my personal favourite involving a seagull. Lorenz wrote: 'one meteorologist remarked that if the theory were correct, one flap of a sea gull's wings would be enough to alter the course of the weather forever. The controversy has not yet been settled, but the most recent evidence seems to favour the sea gulls (Lorenz, 1963).

While it is easy to mock the essential point here is that cause, and effect are not always both obvious and proportionate. Let us try a real example. There is a prominent global cooling event at approximately 8200 years ago that can be traced across multiple climate proxies in different parts of the world. It lasted for only 400 years but reduced global temperature by 6 °C. This event corresponds to a release of a large volume (4.3×10^{14} m³) of meltwater dammed up by the Laurentide (North American) Ice Sheet. When I say large, enough to raise global sea level by 1.2 metres! The most likely cause of the global cooling was a temporary shutdown of the thermohaline conveyor in the Atlantic. All this freshwater would have changed the salinity balance and interrupted the formation of Atlantic Deepwater and therefore the conveyor. A smallish event in the North Atlantic had a profound global consequence.

Another example involves the growth of ice sheets and the role of topography. Let us take two topographic situations. The first is an ice cap forming on the top of a mountain range (Fig. 5.13A). As climate deteriorates more snow falls and the accumulation of compacted snow/ice exceeds melting referred to as ablation and the margins of the ice cap extend sequentially down both side of the mountain range. Climate and ice volume are linked in a linear fashion. Let us contrast this scenario with another in which a large valley glacier descends from a mountain range to spread out over a lowland area as a large piedmont lobe. The Malaspina Glacier in Alaska is a good example of this. The forward velocity of a glacier is a function of the gradient in part, steeper the gradient the faster the flow. So, coming down the side of the mountain range the glacier flows fast, but when it hits the coastal plain it slows down. This is just the same as traffic jam, the cars in front slow and those that are behind are still racing forward and the results if no one breaks is a pile up of cars. The same is true of the ice, it thickens in a non-linear fashion as it enters the piedmont lobe. In fact, it can dramatically accelerate the growth of an ice cap as shown

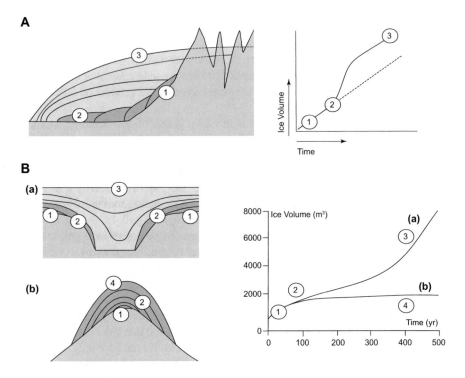

Fig. 5.13 Examples of non-linearity in terms of climate and glacier growth. (Part A is modified from Bennett M.R., Glasser, N.F. 2009. *Glacial Geology,* Wiley, Fig. 3.22. Part B is modified from: Payne, A., Sugden, D., 1990. Topography and ice sheet growth. *Earth Surface Processes and Landforms*15, 625–639, Fig. 6)

in Fig. 5.13B. The ice thickening is out proportion with the climate deterioration and is therefore non-linear. There is another example of this in Fig. 5.13B where glaciers meet in a valley or basin. As the glacier snouts merge melting stops and glacier growth accelerates.

Aside from non-linear responses we also have some simple self-regulatory mechanisms. Take, for example, the link between weathering and temperature. As global temperature rises chemical weathering becomes more effective, which in turn cause a drawdown of carbon dioxide causing global temperatures to fall thereby reducing the loss of atmospheric carbon dioxide. This is a nice example of a planetary scale feedback, or self-regulatory mechanism. It is the sort of regulatory mechanism that James Lovelock believes is evidence of homeostasis on Earth and therefore one reason we should consider it a living organism (Gaia).

Lovelock and his colleagues spent a long time looking for examples of self-regulation and one of the best examples they came up with is known as the CLAW Hypothesis after the authors who discovered it (Charlson, Lovelock, Andreae and Warren). It works something like this. A warming planet cause a bloom of certain phytoplankton, such as coccolithophorids which synthesise something known as dimethylsulfoniopropionate (DMSP). This is released as an osmolyte, basically a process by which of a cell extrudes waste material. This molecule breaks down to give dimethyl sulphide (DMS), first in seawater, and then in the atmosphere. DMS is oxidised in the atmosphere to form sulphur dioxide, and this leads to the production of sulphate aerosols. These aerosols act as cloud condensation nuclei and increase the density of droplets in a cloud and thereby its albedo. As the cloud cover grows more radiation is reflected to space and the temperature cools thereby reducing the bloom of phytoplankton. This feedback loop along with the weather example given before are both examples of negative feedback loops which work to prevent extremes of temperature.

Some feedback loops are positive in sign and therefore increase extremes. Perhaps the most important of these, at least theoretically, is the runaway albedo effect. As temperatures fall snow and ice covers more of the land surface for a greater proportion of the year which increases surface albedo. This in turn causes further cooling, by reflecting more solar radiation, and the snow/ice cover increases. Now computer modelling suggests that if ice cover was to reach 30° north and south of the equator then the process would become unstoppable and the whole planet would freeze. This is the mechanism proposed by some researchers to explain near global glaciations in the late Proterozoic, often referred to as Snowball Episodes.

In summary, there are various mechanisms, many of which are still not well understood or even known, by which small climate changes can be amplified or self-regulated against. It is perhaps now time for us to put this model to the test and explore the climate history of the last 65 million years, that is of the Cenozoic.

Cenozoic Climate Change

So how do we put all this information together? Well probably the best way of looking at this is by focusing on the climate change during the last 65 million years. By focusing on the Cenozoic in this way we can see how all the mechanism and feedback systems we have discussed so far in this chapter work together to bring about change.

The dinosaur bought it in a fiery meteorite strike at the end of the Mesozoic 65 million years ago. The dawn of the age of the mammals was a dark and gloomy affair to start with but soon the warm conditions so typical of the Mesozoic were re-established. All that was missing where those giant reptiles. We can reconstruct the temperature trajectory throughout the Cenozoic, the Era that followed the Mesozoic, by looking at a synthesis of marine sediment cores to create a composite δ18O record (Fig. 5.14).

The first thing to strike me at least when I look at this curve is how from the early Eocene onwards global climate steadily cooled before it became unstable, with rapid glacial-interglacial oscillations, in the Quaternary (last 2.8 million years). We can also see that there were some large tectonic and oceanographic events during this time. We should start our story perhaps with the Early Eocene Climatic

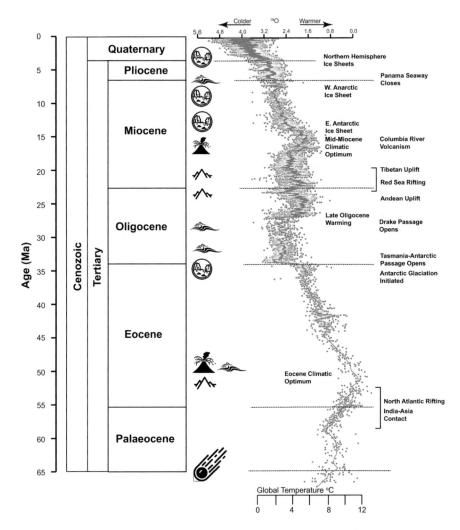

Fig. 5.14 Marine oxygen isotope record for the Cenozoic showing key climatic, tectonic, and oceanographic changes. (Plotted from data in: Zachos, J., et al., 2001. Trends, Rhythms, and Aberrations in Global Climate 65 Ma to Present. *Science*, 292, 686–693)

Optimum, which is often postulated as an analogue for a future greenhouse world. There are two leading hypotheses for the cause of this warm event. The first is the destabilisation of methane clathrate on continental shelves. Now methane clathrate is a molecule of methane within a molecular cage of frozen water, and it forms a solid looking a bit like ice, but unlike ice it will burn when ignited. It forms by the decay of buried organic matter in continental shelf sediments and specific temperature and pressure conditions are needed to keep it in this solid form. Change those temperature and pressure conditions and the clathrate breaks down to give methane which rises through the water column and enters the atmosphere as a greenhouse gas. Remember we said that methane was four times more effective than carbon dioxide as a greenhouse gas. The idea is sometime referred to as a 'runaway greenhouse' in which the release of methane clathrate causes global temperatures to rise and with-it sea level, which in turn cause more methane clathrate to become unstable releasing more methane. Sea level controls water pressure and if it rises pressure at a given depth will increase. The cause of the initial trigger, which sets this feedback off, is not clear, however.

The alternative explanation is linked to the superplume which heralded the opening of the North Atlantic between Greenland and Norway at about this time. If you travel to the isles of Mull or Skye, you will see large horizontal sheets of flood basalt in many of the cliffs. This is part of a huge volcanic province which produced not only large amounts of basalt but also lots of carbon dioxide. This may have caused a rise atmospheric carbon dioxide. Scientists increasingly favour this latter explanation for this climatic optimum. Many marine organisms were challenged at this time, but on land the order of Primates flourished and diversified. It is perhaps slightly ironic to think that ultimate success of primates (and ultimately our species) may be linked to period of greenhouse warming.

The progressive drawdown of carbon dioxide and the slow decent toward an icehouse world continued throughout the Eocene. The cause was the progressive growth of the Himalaya and in time the Tibetan Plateau. This not only exposed huge quantities of fresh bedrock for weathering but also altered the cause of the Rossby Waves in the upper atmosphere. Remember that weathering scrubs carbon dioxide from the atmosphere and the less carbon dioxide the colder our planet will become.

It was the separation of Tasmania from Antarctica that led to the next step and the initiation of glaciation in Antarctica (Fig. 5.15). The opening of the sea way was the

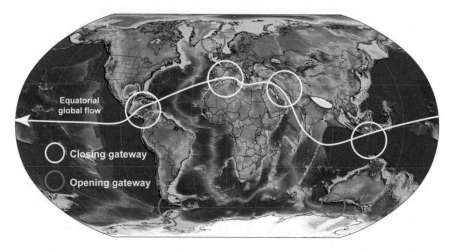

Fig. 5.15 Summary of key oceanographic events in the Late Eocene around 35 million years ago. (Palaeogeographic map reproduced with permission from Chris Scotese. See also: Scotese, C.R., 2021. An Atlas of Phanerozoic Paleogeographic Maps: The Seas Come In and the Seas Go Out. *Annual Review of Earth and Planetary Sciences* 49, 669–718)

start of process that would see Antarctica separated from the southern mid-latitudes by a circum-polar current. The sudden drop in global temperature at the Eocene to Oligocene boundary reflects the presence of ice in Antarctica and the associated increase in albedo. This cooling caused a minor extinction event with primates wiped out from North America.

The Drake Passage which separates South America from the Antarctic Peninsula opened in the middle of the Oligocene and completed the isolation of Antarctica. The Antarctic Circumpolar Current was created, and it is an effective barrier to the movement of warm water (and air) from mid-latitudes and led to the progressive grip of ice on the Antarctic Continent. The Oligocene and early part of the Miocene was a period of relative stability including a brief swing back to warmer conditions caused perhaps by the release of carbon dioxide. This warm hiatus corresponds in age to the La Garita Caldera in Colorado. This volcanic eruption is one of the largest known volcanic eruptions in Earth's history. The deposit, known as the Fish Canyon Tuff, covered at least 28,000 km^2 with an average thickness of over 100 m. The climate optimum in the Miocene ten million or so years later may also be linked to a volcanic event this time known as the Columbia River Volcanic Province. It is the youngest, smallest and one of the best-preserved continental flood basalt provinces on Earth and covers over 210,000 km^2 in mainly in the states of Oregon and Washington.

Throughout the Miocene continued reorganisation of ocean currents continued with the closure of key gateways, creating the segmented pattern of currents we see today (Fig. 5.15). Around this time the lip between the North Atlantic and the Artic Basin, referred to the Greenland-Scotland Ridge began to subside allowing cold water to flow from the Arctic into the Atlantic strengthening the thermohaline conveyor and configuring it as we see it today. The re-position of North America and Asia into northern mid-latitudes continued at this time. It has been suggested that the major plate tectonic re-organisation which led to all of these changes was linked to the activity of two giant superplumes originating deep in the mantle. Both direct volcanism and more general upwelling around these superplumes may have facilitated this re-organisation. Some of the key plate tectonic events in terms of climate are summarised in Table 5.1. By the middle of the Miocene the Earth had firmly shifted towards becoming an icehouse.

The development of the East Antarctic Ice Sheet around 12 Ma and subsequently the West Antarctic Ice Sheet at the start of the Pliocene saw further step wise cooling against a background of progressive cooling (Fig. 5.14). By the start of the Quaternary at 2.8 Ma years ago planet Earth was firmly in the grip of an ice age. The ocean and atmospheric systems were in a critical state of instability and primed so that the subtle orbital variations in solar input could drive rapid swings in climate from glacial to interglacial.

Milanković cycles have always operated, the gravitational pull of Saturn and Jupiter are not new after all, but they now were amplified sufficiently to drive major swings in global environment. The world was plunged into a series of short interglacial separated by longer glacials. The amplifying mechanism is still debated but is likely to involve the thermohaline conveyor.

Table 5.1 Key gateway oceanographic events in the Cenozoic

	Gateway	Timing	Significance
Opening	Tasmanian gateway between Australia and Antarctica	33–35 Ma	Formation of the Antarctic circumpolar current (ACC), which isolated Antarctica from warmer waters; early Antarctic ice sheets form
	Drake Passage between South America and Antarctica	41 Ma	Beginning of the present Antarctic circumpolar current–key player in deep-water global circulation–and colder global climate
	Greenland–Scotland ridge	12 Ma	Arctic cold water enters Atlantic. Before subsidence, waters of Arctic Ocean isolated from world ocean
Closing	Tethyan Ocean in Middle East	15–17 Ma	Initial closing of connection between Pacific and Atlantic oceans leading to isolation of Mediterranean and Black seas coincides with increased aridity in Caspian–Black Sea region and Middle East
	Indonesian gateway between Borneo and New Guinea	Started 25 Ma, closed by early Miocene	Closed to deep flow between Pacific and Indian oceans export of heat from Pacific to Indian Ocean may have affected monsoon system of Southeast Asia
	Panamanian isthmus between central and South America	Initial 15 Ma, final closure at 3.5 Ma	Culmination of 13 Ma restrictions beginning in the Miocene. Greatly strengthened gulf stream and had strong effects on marine biotas and on vertebrates living in central and South America.
	Gibraltar	Shallowing and narrowing in the Neogene with brief closure at end of the Miocene	Brief closure leads to Mediterranean drawdown and salt plus canyons and gorges of rivers such as those of the Nile

Based on information in: Potter, P.E., Szatmari, P., 2009. Global Miocene tectonics and the modern world. *Earth-Science Reviews* 96, 279–295

Milanković cycles have most impact in the northern mid-latitudes where in the Quaternary the largest concentration of land was located. Changing both the absolute amount of radiation received and its seasonal distribution may have been sufficiently to allow winter snowfall to at first remain longer each year before becoming permanent. More snow cover would also lead to more cooling via albedo feedback. Changing patterns of runoff from mountains around the North Atlantic and ultimately from the ice sheets that developed would impact on the salinity of the North Atlantic and may have caused periodic weaken of the thermohaline conveyor amplify the subtle radiation variations around the globe. The prime mechanisms are not entirely clear and subject ongoing research, but the coupling of local radiation balance, ice sheets and oceans is key to the story. Some researchers have suggested that the conveyor may operate in three modes (Fig. 5.16) and when it is turned off

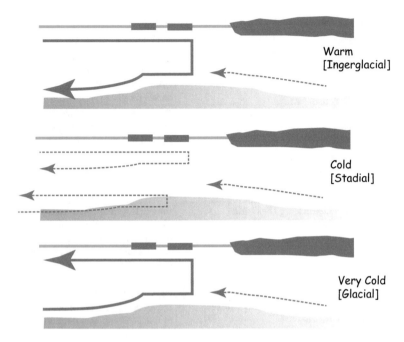

Fig. 5.16 Three possible modes of the thermohaline conveyor in the North Atlantic

the area around the North Atlantic is plunged into a glacial. Why the amplifier responded to different Milanković cycle at different times is not entirely clear with various competing theories being advocated by different research groups.

We have a basic model for climate change during the Cenozoic the gradually repositioning of the continents via plate tectonics, coupled with mountain building and the isolation of Antarctica. This all created a critical climate state in which small variation in the amount and seasonal distribution of radiation received in the norther mid-latitudes as result of orbital cycles were sufficient to cause global environmental swings between glacial and interglacial states. The thermohaline conveyor may have played an important role in amplifying these subtle variations and ensuring that environmental response was synchronous more or less around the globe.

It is at this point that we need to spoil our nice story with a few more facts. These facts largely came from the ice core records of Greenland. These records provided evidence of lots of climate change events at sub- Milanković time scales; events that are simply too short in length to be explained by orbital variations. One of the most striking features of the ice core record is the rapidity of some climate events, with temperatures warming by 1 °C every five years in some cases. These rapid events also mirror the shape of the Milanković events (Figs. 5.4 and 5.17) which have a saw-tooth form in which slow progress cooling is followed by rapid warming.

This shape is not predicted by Milanković cycles. The rapid events identified in the ice cores also have this shape which may speak to a common cause. The cycles within a Glacial are called Bond Cycles (Fig. 5.17) and within a Bond Cycle we

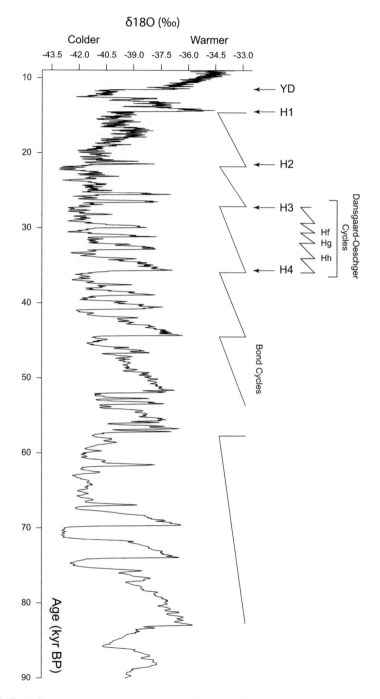

Fig. 5.17 GRIP ice core record. Note the saw-tooth shape within this single Glacial episode. Bond Cycles are groups of Dansgaard-Oeschger cycles as illustrated. Also note the peaks in lithic content on the ocean floor, so called Heinrich Events. (Plotted from data in: Johnsen, S. J. et al., 1997. The d18O record along the Greenland Ice Core Project deep ice core and the problem of possible Eemian climatic instability. *Journal of Geophysical Research: Oceans* 102, 26,397–26,410. Available from: https://www.ncdc.noaa.gov/paleo-search/study/17839)

have Dansgaard-Oeschger (D-O) Cycles all record slow cooling followed by rapid warming. The warming at the end of a Bond Cycle is proceeded by what has become known as a Heinrich Event. Yes, I know everyone wants to get in on the naming of a climate event!

Heinrich Events where first found in ocean cores in the North Atlantic and consist of coarser horizons. These horizons result from debris settling out from lots of icebergs, people talk about armadas of icebergs. By looking at the rock type and minerology of these layers you can trace them back to the Hudson Bay region. Lots of icebergs were released into the Atlantic from ice sheets in Canada, as they spread out over the ocean, they melted releasing their debris as an event horizon. These events occur just before a rapid warming at coldest point in one of these cycles. Figure 5.17 shows a section of ice core record with the Bond Cycles and Heinrich Events marked. Smaller Heinrich Events also occur at the end of some of the D-O Cycles. All that freshwater (icebergs are from glaciers and are fresh) added to the Atlantic would cause the Conveyor to stop briefly and perhaps then re-set itself. This forms one of the possible explanations for these cycles as shown in the cartoon in Fig. 5.18.

The idea is that to start with the ice sheets are quite small and do not reach to the edge of the continental shelf. Corridors of fast flow know as ice streams are switched

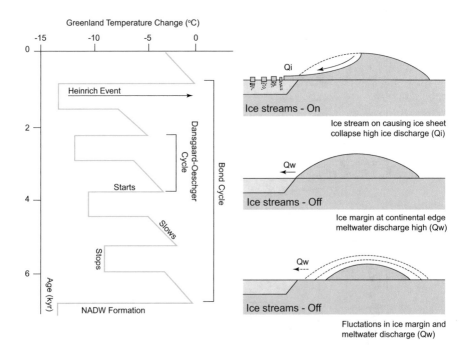

Fig. 5.18 Cartoon showing the relationship between D-O and Bond Cycles each controlled by the strength of the thermohaline conveyor and terminated by a Heinrich Event in which the ice sheets are so big that they reach the edge of the continental shelf and become unstable. (Modified from Alley, R.B., 1998. Icing the north Atlantic. *Nature* 392, 335–337, Fig. 1)

off. As the ice sheets grow progressively with falling temperature meltwater inputs to the ocean increase and begin to cause fluctuation in the strength of the thermohaline conveyor, small stutters if you like. These are the D-O Cycles. However, as the ice sheet grows bigger it eventually reaches the edge of the continental shelf and or become increasingly unstable. Ice streams switch on and lead to the rapid discharge of ice to calving margins on the shelf edge creating large armadas of icebergs. All this freshwater causes the thermohaline conveyor to stall or stop. The purging of ice leaves the ice sheets much smaller and removes their influence on ocean and conveyor which rapidly re-sets into a warm interglacial state. Effectively we have a binge-purge cycle operating within the ice sheets. They binge on ice and grow to the limits of the continental area, or pass some internal stability threshold, and then purge themselves through rapid ice stream flow.

The reality is probably something a bit more complex than this but what is nice about this model is that ice sheet stability provides a check on the glacial cycle. Their inherent instability pulls back the system to warmth. If you are interested there are many different ideas at play here and it is an active area of research in which new theories emerge all the time.

There is one final piece of this puzzle that is worth spending a few moments on. The last glacial cycle, called the Devensian in the UK, reached its peak around 18,000 years ago in what is known as the Last Glacial Maximum or LGM. This is when glaciers were at their most extensive in Britain, Fennoscandinavia, and in North America (Laurentide Ice Sheet). Milanković does a good job at predicting the timing of this event and off the subsequently warming. Ice sheets were retreating fast and much of the UK at least was ice free when suddenly things got cold again and quite suddenly. This stadial as it is called is not predicted by Milanković in fact solar radiation was peaking. So why was there a short, sharp return to glacial conditions?

This event is known as the Younger Dryas (Stadial) and occurred between about 12,900 and 11,700 years ago. In the UK we call the event the Loch Lomond Readvance, because glaciers returned to many upland areas and in Scotland retreating remnants of the last ice sheet began to advance again. The glaciers of the Younger Dryas were the last to exist in Britain's uplands and therefore the evidence is both fresh and easily mapped. For decades researchers have argued over these deposits and used them to reconstruct the small ice caps and glaciers that existed at this time. So, in short British glacial geologists are a bit obsessed with this event!

The prevailing idea is that it was caused by a brief shutdown of the thermohaline conveyor due to the release of huge amounts of fresh water into the Atlantic. To understand this, we need to think about the Laurentide Ice Sheet (North American Ice Sheet). Located mainly in Canada its southern margin extended to the Great Lakes. The St Lawrence Seaway was blocked by ice. Now an ice sheet some three thousand metres thick is quite heavy and will cause the crust beneath it subside due to isostasy (Fig. 2.5), but because the crust is quite strong the area of depression extends beyond the limits of the ice sheet by quite a distance. In fact, we get a small forebulge several hundred kilometres in front of the ice sheet margin and a depression between this bulge and the ice sheet edge. In the case of the southern margin of the Laurentide Ice Sheet this depressed area just in front of the margin corresponds to the Great Lakes. The result is that we get supersized Great Lakes, which drain the only way they can which is south into the Mississippi River and onwards to the Gulf of Mexico. Now that is a lot of freshwaters, and it is kept from the Atlantic by the ice blocking the St Lawrence Seaway. Provided the weight of the ice is greater than the pressure of the water all is stable, but as the ice thinned after the LGM its weight reduced yet the water depth in front increased. Crust is slow to adjust so remains depressed. Ultimately this will become unstable and at first a trickle of water found its way beneath the ice dam, but that trickle generates frictional heat, which melts the ice and before you know it you have a cataclysmic flood beneath ice sheet and into the Atlantic. All this freshwater shut the thermohaline conveyor down and plunge the Atlantic region back into full glacial conditions for a brief period. It is believed that the same thing happened at 8200 years ago causing another, although briefer, cold spell.

- Task 5.5: What can you find out about the 8200-climate event? Was it global? What sort of things record the deterioration in climate, be precise and find examples?

It is a nice theory and for many researchers has a certain degree of elegancy about it. It may not be the entire story, however. Recently some researchers have begun to advocate and extra-terrestrial cause for the Younger Dryas in the form of a series of meteor strikes. The recent discovery of a large crater beneath the Greenland Ice Sheet has given some momentum to this idea, although it is hotly debated.

Summary

For some understanding climate change in the recent past may be challenging. I agree that some of the concepts are complex and there is no simple answer. Different mechanisms nest one within another. It is also an area of active research and new evidence is constantly emerging leading to new ideas. It is worth persevering with because unless you can understand the climate system and how it works and changes by natural process you cannot really advocate effectively, at least from an informed position, about the climate challenges that face us in the coming years. Climate change is not a simple human caused phenomena, climate changes naturally and rapidly all on its own!

To help summarise all the different nested mechanisms of climate change I have produced a summary diagram (Fig. 5.19) which has everything on it that you need to know.

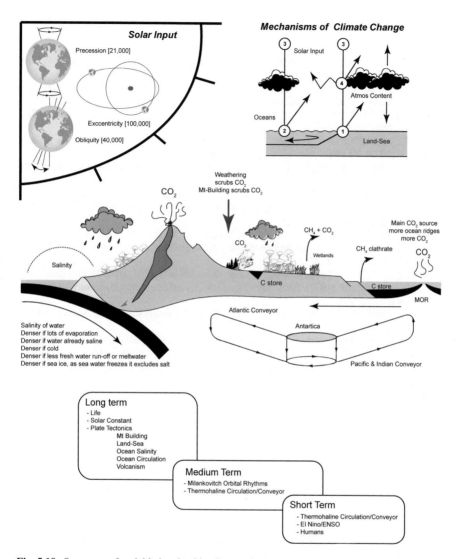

Fig. 5.19 Summary of variable involved in climate change

Further Reading[1]

Alley, R. B. (2015). *The two-mile time machine: Ice cores, abrupt climate change, and our future.* Princeton University Press.

Atkinson, T. C., Briffa, K. R., & Coope, G. R. (1987). Seasonal temperatures in Britain during the past 22,000 years, reconstructed using beetle remains. *Nature, 325*, 587–592.

Barriopedro, D., et al. (2014). Witnessing North Atlantic westerlies variability from ships' logbooks (1685–2008). *Climate Dynamics, 43*, 939–955.

Cockell, C., et al. (2008). *An introduction to the earth-life system.* Cambridge University Press.

Feulner, G. (2012). The faint young sun problem. *Reviews of Geophysics, 50*, RG2006. / 2012.

Hays, J. D., Imbrie, J., & Shackleton, N. J. (1976). Variations in the Earth's orbit: Pacemaker of the ice ages. *Science, 194*, 1121–1132.

Imbrie, J., & Imbrie, K. P. (1986). *Ice ages: Solving the mystery.* Harvard University Press.

Lorenz, E. N. (1963). The predictability of hydrodynamic flow. *Transactions of the New York Academy of Sciences., 25*, 409–432.

Maslin, M. (2016). Forty years of linking orbits to ice ages. *Nature, 540*, 208–209.

Urey, H. C. (1947). The thermodynamic properties of isotopic substances. *Journal of the Chemical Society (Resumed)*, 562–581.

Ruddiman, W. F. (2017). *Plows, plagues, and petroleum.* Princeton University Press.

Sagan, C., & Mullen, G. (1972). Earth and Mars: Evolution of atmospheres and surface temperatures. *Science, 177*, 52–56.

Smithson, P., Addison, K., & Atkinson, K. (2013). *Fundamentals of the physical environment* (4th ed.). Routledge.

Wheeler, D., et al. (2009). Reconstructing the trajectory of the august 1680 hurricane from contemporary records. *Bulletin of the American Meteorological Society, 90*, 971–978.

Wilson, R. C. L., et al. (2000). *The great ice age: Climate change and life.* Routledge.

---

[1] Most physical geography textbooks contain a chapter on climate change and will discuss at the very least Milanković cycles. For example, Smithson et al. (2013) has some material although is not as good on this subject as one might hope. Charles Cockell and colleagues gives a more advanced account in *An Introduction to the Earth-Life System* that I think is particularly good. There is also another Open University text called *The Great Ice Age* by Wilson et al. (2000) which is extremely good if a little detailed for the level. Do not be put off that it is a bit old, it is a great book!

There are some fantastic American primers on aspects of this subject. *The Two-Mile Time Machine,* by Richard Alley, revised in 2015 and published by Princeton University Press is simply brilliant and a great read. It deals with the US Greenland Ice Cores. *Plows, Plagues, and Petroleum: How Humans Took Control of Climate* by Bill Ruddiman and revised in 2017 again published by Princeton University Press is a thought provoking read and likely to challenge some of your perspectives on climate change. Imbrie and Imbrie (1986) give an account of the Milanković ice age theory and the role that oxygen isotopes played in its vindication. There is a paper in *Nature* by Mark Maslin (2016) which provides a retrospective on Milanković cycles and has some good information in it and is accessible.

Part II
Shaping the Landscape

Chapter 6
The Landscape Beneath Our Feet

I love landscapes they always have a story to tell and are themselves the setting for the best stories. Clues in a landscape allow us to evidence stories about the past and the evolution of that landscape over time. There are forces which build up the land, tectonic and volcanic ones from within the Earth(endogenous) and those which are external (exogenous), the agents of entropy, that work to wear it down. Between these forces we have the rocks on which they act, and rock type (lithology) is an important if somewhat neglected variable in the study of landscape evolution. It is one which is key to the landscape character of different parts of the British Isles for example. Take a look at the cross section shown in Fig. 5.1 which runs from Snowdonia (North Wales) to the English Channel. It is based on a famous cross-section drawn first by William Smith (1769–1839) as part of a nationwide geological map, the first such map of any country in the World. He championed the correlation of rock units by the use of fossils and when his map was first published it was overlooked by the scientific community due to his humble education and lack of family connections. In later life however he was duly recognised as one of the "Fathers of English Geology" according to Winchester (2002).

On this cross-section contrast the landscape of the Wealden Anticline in the Southeast with the rugged mountains of Snowdonia. Geology truly defines the landscapes and the flora/fauna of Britain (Fig. 6.1). If by some wild chance you are less interested in ecology and more interested in Roman history take a look at Fig. 6.2. Here the boundaries of 'Romanisation' of Britain are shown and to my eye at least they are highly influenced by terrain. In fact, terrain limited the Roman expansion into Wales and the Roman frontiers follow the geological boundaries (roughly) and terrain identified in the cross section in Fig. 6.1. The influence of geology on history and warfare is an interesting muse and there are many research projects in this area waiting to be done.

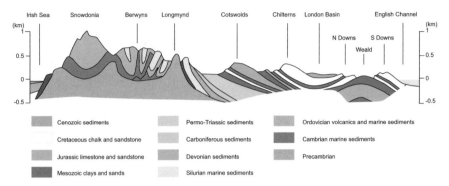

Fig. 6.1 Geological cross-section showing the main rock units (+ their relative age) along a line between Snowdonia and the English Channel. (Inspired by William's Smith famous map and cross-section but re-drawn from Smithson, P., Addison, K., Atkinson, K., 2008. Fundamentals of the physical environment. Third edition Routledge, Figure 13.6)

To understand these fundamental controls on ecology and history, we need to know more about different types of rocks and their resistance to weathering and erosion.

- Task 6.1: Using Google Earth follow the cross-section in Fig. 6.1. How does the landscape vary along this line? Now take a look at Areas of Outstanding Natural Beauty (AONBs) and National Parks in the UK, there are listings to be found on the internet. Pick two or three along the line of cross-section and consider their different ecologies. Think about the importance of geology and terrain as the controlling variable.

Rocks and Minerals

Our planet is a silicon (Si) dominated one. Most although not all rock-forming minerals are silicon based and rocks are the building blocks of our planet. About 75% of all rocks are made of silicon and oxygen and there are over 2000 minerals. Around 80% of the Earth's surface is composed of igneous and metamorphic rocks with the remainder made of sedimentary rocks. Only 5% of planet is composed of non-silicate-based minerals, mainly carbonates and evaporites. The core block is a silicate tetrahedron that is one silicon atom surrounded by four oxygen atoms. These tetrahedra can occur on their own or be organised into various crystalline structures. In a tetrahedron there is a net charge of −4, which enables other tetrahedra to link via a shared oxygen atom or bind with other cations. Octahedra is another possible configuration with six oxygen atoms around one silicon.

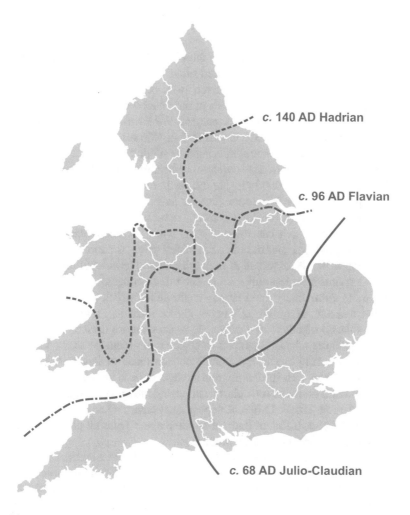

Fig. 6.2 Romanisation of Britain, consider how these boundaries reflect the geology of the Britain and the terrain picked out in the geological cross section in Fig. 6.1. (Source: https://en.wikipedia. org/wiki/Roman_conquest_of_Britain)

In this way we can create complex arrangements of tetrahedra and octahedra to form crystal latices such as ring silicates, chain silicates and sheet silicates. We can add to this diversity by substituting silicon for aluminium and joining these silica building blocks with other atoms including iron, sodium, magnesium, calcium, and titanium. This is why we end up with a lot of different rock forming minerals. The crystal structure of a mineral controls the way light passes through it and therefore its colour and other optical properties. The structure also controls its strength. We then have the carbonate minerals of calcite and aragonite which are found in bones and shells. Rocks made of the carbonate remains of animals gives us limestone and chalk. We can also evaporite salts to give us halite and gypsum which are largely silicon free.

As we saw in Fig. 2.2 rocks fall into three broad categories, igneous, metamorphic, and sedimentary rocks. All contain minerals but in the case of sedimentary rocks it is the mineral fragments derived from other rocks that make them up. Igneous rocks are classified on the basis of the mineral crystal size and the chemistry, specifically the amount of silica present. We have a crude continuum which starts with ultramafic rocks which have less than 40% silica dioxide in them and are composed almost entirely of minerals such as olivine. The continuum ends with acidic rocks with such minerals as quartz, orthoclase feldspar, albite, and muscovite. If acidic rocks cool slowly inside the earth, then we get granite, if they cool rapidly as part of a volcanic eruption then we have rhyolite. At the other end of the spectrum the most mafic of rocks is peridotite which occurs in the mantle. Basalt is another mafic rock which has small crystals since it cools quickly in a volcanic eruption. Cooled slowly and the same chemistry gives us gabbro. Crystal size and mineral composition is important to rock strength and its susceptibility to weathering. Small crystals have less surface area and often stronger.

We classify sedimentary rocks largely on the characteristics of the grains or fragments that make them up. We talk about clastic rocks as those derived from the fragments (clasts) of others and carbonate rocks as those derived from the fossil remains of other organisms. Examples of clastic rocks include sandstones, made of sand, siltstones, made of silt, and conglomerates made or larger rock fragments (pebbles and cobbles). The properties and strength depend on two things the properties of the mineral grains or rock clasts composing the the rock, their size and sorting (range of grain sizes). Different depositional environments deserts, rivers and the sea give the rock different internal structures and layering patterns (bedding) which also help determine a rock's properties.

Metamorphic rocks are rocks that have been transformed by the application of heat and pressure. This causes the minerals to re-form and also to become orientated in response to the applied pressure. The types of minerals and therefore the type of metamorphic rock is a function of heat and pressure and also the starting material. For example, marble is transformed from limestone, slate from siltstone and so on. We talk about the metamorphic grade the higher the grade the more intense the heat and pressure applied.

- Task 6.2: Pick five rocks that you like the sound off and research their properties. What are they composed off where do they form and how can you distinguish them from other rocks?

Natural Decay

Entropy requires a rock to decay, and it will over time through the process of weathering. Weathering is the *in-situ* breakdown of rock whereas erosion is the breakdown *and* transport. Semantics yes but useful to distinguish between the two processes. We can recognise three families of weathering process mechanical (physical), chemical and biological. Physical weathering involves breaking a rock down into smaller fragments effectively increasing the surface area. Take a cube and divide it into eight smaller ones and then 16 and so on. Each time we divide the cube you increase the surface area exponentially. Why does this matter? Well chemical process operates on a surface so by increasing the surface area you increase the potential for chemical weathering to operate. Physical weathering therefore helps accelerate the action of other weathering processes. Chemical weathering involves the alteration of the rock forming minerals present. Some minerals are resistant to weathering like quartz and will tend to remain as residual minerals. Other minerals are transformed chemically and structurally into sheet silicates known as clay minerals. Also, some material is lost in solution.

Perhaps the most potent of physical weathering processes is down to salt. A humble kettle builds up limescale in hardwater areas such as Bournemouth. The constant evaporation of water leads to its the precipitation around the filament and in the spout. Groundwater contains salts especially in semi-arid regions and as groundwater moves through permeable rock the salts within it are precipitated by evaporation and build up in the pores, voids, and fractures. The more porous a rock the more voids there is for this to occur. Constant wetting and drying helps this process. As the salts build up, they exert a pressure on a rock, ultimately pushing mineral grains out or helping to open fractures. It is effective, very effective in fact and has been the subject of lots of experimentation. It is relatively easy to take a block of rock of known mass and subject it to wetting and drying cycles in a lab using different salt solutions. Sodium sulphate has been found to be one of the most effective salts in driving this process and rock porosity is also important with limestone weathering quickly, whereas dolerite, which is impermeable, remains impervious at least in the short term. The other aspect of salt weathering worth noting is that the salts will expand under heating at different rates than the rock around them and this can also exert a destructive force over time.

- Task 6.3: One of the classic papers on salt weathering is Cooke (1979) check it out and try to make some notes.

Another commonly discussed physical weathering process is freeze-thaw. Anyone with an uninsulated water pipe will know about this. As a pipe freezes the water expands by 9% and because this expansion is constrained by the pipe it will often fracture. So, in theory water in a porous rock, or in a fracture of any rock, will expand on freezing causing pressure to be applied. Repeat this over time and you will get failure due to repetitive stress. This sounds plausible, but the problem is that a pipe is a closed system in which the pressure can build up, while a fracture in a rock is not. The ice simply expands out of the crack and the lateral forces are much reduced. Yet go to any regions with a regular freeze-thaw cycle and you can see the rock destruction for yourself (Fig. 6.3).

The truth is that freeze-thaw weathering is surprisingly complex and despite a huge amount of research remains a bit of mystery. The movement of super-cooled water within rock pores and fractures towards a freezing front that moves through a rock (surface inwards) is one potentially important factor. This is how lens of segregated ice develop in soils; the water migrates and concentrates where it is already frozen. Thermal shock is another factor that might come into play. All I can say here is that research continues but go to any polar region and the evidence for the efficiency of the process is there to be seen.

- Task 6.4: Check out the paper by Nicholson and Nicholson (2000) paper which discusses experimental results for freeze-thaw weathering of different rock types. It shows you how simple lab experiments can be used to study weathering.

Different materials expand and contract at different rates so in theory different minerals should also expand and contract differently and this has been proposed as another physical weathering process. A variant on this is due to surface heating of rock; the surface heats quickly and expands but internally the rock remains cool due to the poor thermal conductivity of most rocks. The stresses set up in this way may in time cause the rock to flake with surface layers breaking free. Again, there is a lot of experimentation around this process enhanced more recently by direct monitoring of rock temperature. Despite this there is also a lack of precise clarity about how some of these processes actually work. The results of physical weathering look like

Fig. 6.3 The results of freeze thaw in action in Svalbard. Note how in B fracturing is concentrated in certain layers of the clast centre frame. These are more porous layers

the result of a giant wielding a sledgehammer, but this betrays a complexity which is still at the limits of our knowledge.

In terms of chemical weathering, we have four main process suites one is oxidation/reduction, another is solution/hydrolysis, then we have hydration and isomorphous replacement. Take the element iron when oxidised it has the colour of rust and is insoluble but in the absence of air, that is in its reduced state, it is blue in colour and soluble in water. A range of minerals and elements have different redox states which effect their properties and susceptibility to weathering. Solution occurs when a soluble element is removed from a mineral by a solute, usually water, in most natural settings. If this water is acidic then the process is hydrolysis. Rainwater is often acidic naturally since rain or water percolating through soil tends to dissolve carbon dioxide and become weakly acidic.

Solution rates are controlled by the aggressiveness of the water determined by its acidity and degree of saturation with respect to the element concerned. We also need to consider the residence time. Let us think of an empty bus; it has a large transport capacity initially. At each bus stop it fills up until it reaches capacity. The bus will fill quickly if there are lots of bus stops with people and it stops at them all. A friendly encouraging conductor may also help, but even so as it fills some may choose to wait for the next less crowded bus. Once at capacity the bus must first discharging passengers before it can continue to transport new ones. Water is the bus; the passengers are the soluble components; the conduct's Bon amie and bus's size equate crudely to the aggressiveness of the transport process.

If water sits on the mineral surface and does not move, then it will quickly become saturated with that element and solution will cease. Equally if the water is flowing to quickly then it never has time to reach its full carrying capacity (maximum concentration). Temperature also affects the speed of the reaction as does the acidity of the water. There is a whole area of research which deals with solution of carbonate rocks, limestone and chalk, and such landscape are referred to as karst are direct products of weathering and associated fluvial erosion by solution. Britain and Ireland have extensive areas of limestone of Carboniferous age and karst terrain is a particular feature in parts of Northern England for example.

- Task 6.5: Limestone pavement (Fig. 6.4) is a feature of karst terrain in Britain and Ireland. Its formation requires a particular combination of glacial erosion followed by solution weathering/erosion. It is the only geomorphological landform in Britain to have its own conservation legislation (Limestone Pavement Order, Natural England, 2020). Find out more about limestone pavement its morphology, its unique flora and conservation status.

Isomorphous replacement involves the substitution of one element in a crystal lattice, usually silicon, for another. While chemically the same the size of the new atom may be different causing the lattice to become stressed. In principle the same is true with hydration where the addition of water molecules causes the lattice to be altered and/or stressed.

Chemical weathering causes the transformation of a mineral by both changing it chemically and causing the existing crystal lattice to become stressed and ultimately

Fig. 6.4 Limestone pavement in Yorkshire. Note the clints and grykes and the vegetation with its own microclimate in the grykes. (Reproduced with permission from Shutterstock, Totajla)

breakdown. Changing the lattice and the components is a key part in the creation of secondary minerals such as clay minerals. Clay minerals are a type of sheet silicate made up of layers (platelets) held together by electrical charges.

In terms of completing our review of weathering processes we need to mention biological agents. In many respects they are not really any different from physical and chemical processes although a biological agent is at play. Tree roots are perhaps the simplest of these (Figs. 6.5 and 6.6). The roots of a sapling find their way into a rock fracture and as the tree grows and the root expands it acts as a wedge increasing the size of the fracture. In practice however the process also involves chemical factors. In particular the Mycorrhizas which are beneficial fungi growing in association with plant roots influence chemical processes. These fungi exist by taking sugars from plants in exchange for moisture and nutrients gathered from the soil by the fungal strands. They effectively increase the surface areas of the root system. They contribute to weathering by extruding acids and organic chelators which are compounds that combine with metals to make them mobile. They feed a wider range of bacteria and fungi which may accelerate mineral weathering, while also increasing the concentration of carbon dioxide in soil and thereby adding to the acidity of soil water.

Lichens are another effective weathering tool this involves a combination of both mechanical and chemical processes. Generally, weathering of rocks by lichens proceeds by: (1) the penetration of hyphae through intergranular voids and mineral cleavage planes; (2) expansion and contraction of thallus (body) by microclimatic

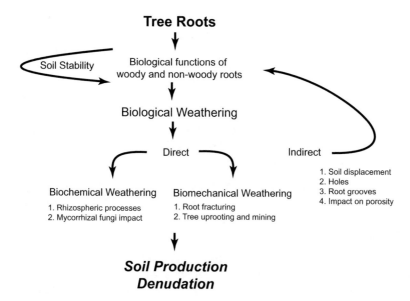

Tree Roots

Soil Stability

Biological functions of
woody and non-woody roots

Biological Weathering

Direct

Indirect

Biochemical Weathering

1. Rhizospheric processes
2. Mycorrhizal fungi impact

Biomechanical Weathering

1. Root fracturing
2. Tree uprooting and mining

1. Soil displacement
2. Holes
3. Root grooves
4. Impact on porosity

*Soil Production
Denudation*

Fig. 6.5 Model of weathering by tree roots. (Pawlik, Ł., Phillips, J.D., Šamonil, P., 2016. Roots, rock, and regolith: Biomechanical and biochemical weathering by trees and its impact on hill-slopes—A critical literature review. *Earth-Science Reviews* 159, 142–159, Figure 13)

Fig. 6.6 Tree roots fracturing rock. (Reproduced with permission from Shutterstock, Drop Zone Drone)

wetting and drying; (3) freezing and thawing of lichen thallus and associated micro-environment; (4) swelling action of organic and inorganic salts originating from lichen activity, and (5) incorporation of mineral fragments into the thallus dissolution of respiratory CO_2 in water held by lichen thalli results in the generation of carbonic acid. Oxalic acid secreted by the mycobionts of many lichens is also commonly considered to play an important role in the chemical weathering of the rock on which the lichen sits.

The occurrence of lichens may date from the Early Devonian (400 million years ago) and may have been one of the first colonizers of terrestrial habitats on the Earth. Today they cover as much as 8% of the Earth's surface.

- Task 6.6: There is a detailed review paper on lichen weathering written by Chen et al. (2000) which you might find interesting. If you like lichens, why not check it out? If you prefer trees, then the review by Pawlik et al. (2016) might be of more interest.

Now whatever the process of weathering temperature and rainfall, climate in other words, is going to have a role in determining what process operate where, and with what efficiency. There is a commonly reproduced diagram which shows a profile from high latitudes to low of depth to rockhead. Rockhead is the junction between soil/regolith and rock. Regolith is the soil.

In the tropics the rockhead is deep and the weathered regolith and soil above thick. This is to be expected since the warm and moist conditions favour rapid chemical weathering. As we move to the subtropical deserts the rockhead approaches the surface and we only get a thin layer of regolith. Here physical weathering dominates, and the lack of moisture tends to limit chemical weathering. Moving to temperate mid-latitudes the increased moisture allows weathering to increase and the depth to the rockhead increases but not too the depths in the tropics. Here it is temperature that limits weathering. In polar latitudes then mechanical weathering dominates once again and the rockhead is near surface. The graph in Fig. 6.7 summaries this the two on the abscissa (x-axis) we have rainfall and, on the ordinate (y-axis) we have temperature between which various fields of process dominance are defined.

Rock type is clearly important, different minerals have different susceptibility to weathering. The more calcium, magnesium, and iron there is in a mineral the more susceptible it is too weathering. Quartz is one of the most resistant being made of pure silica, while olivine is more susceptible. In terms of landforms that are the direct product of weathering then karst scenery is clearly one. The other are tors which dominate the moors of Dartmoor and Exmoor in the West Country of Britain (Fig. 6.8). The classic model of tor formation developed by David Linton in 1955 involves a two-fold process in which deep chemical weathering of granite occurs concentrated in those areas that are densely jointed. The classic model suggest that this occurs in Tertiary before the Quaternary ice age. Then during a second phase this weathered regolith is stripped away by periglacial processes during the Quaternary to leave core stone of un-weathered rock upstanding as tors. A small roadside quarry at Two Bridges not far from Dartmoor Prison is the classic site

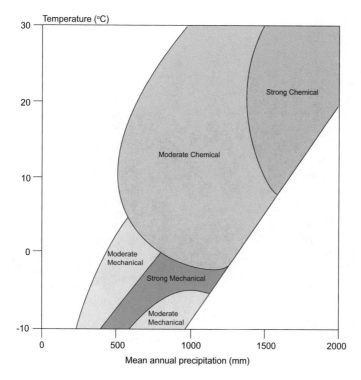

Fig. 6.7 Conceptual model of weathering zones in relation to temperature and rainfall. (Simplified from Peltier, L.C., 1950. The geographic cycle in periglacial regions as it is related to climatic geomorphology. *Annals of the association of American Geographers* 40, 214–236. Figures 1 to 6. See also: Smithson, P., Addison, K., Atkinson, K., 2008. Fundamentals of the physical environment. Third edition Routledge, Figure 13.7)

which supposedly illustrates this model. The model uses a key geomorphological tool known as ergodic reasoning which is basically time-space substitution. Now this all sound a bit like an episode of Star Trek but think about it like this. If we were aliens and wanted to study how humans grow, allometry is the posh name, then we could abduct one and watch it grow up in a laboratory. Alternatively, we could simply go to a shopping mall and measure everyone there, effectively this spatial sample would be a temporal one because people of all ages would be present. That is what ergodic reasoning is; if we have landforms in various stages of formation, we can organise them into an evolutionary timeline youngest to oldest. This is what Linton did with tors and the Two Bridges roadside quarry represented a nascent example, while Hay Tor is an example of a mature form. There is not much alternative since dating erosional forms is difficult or was until recently.

In the last decade this has changed with the emergence and validation of cosmogenic exposure dating. The Earth is bombarded with primary cosmic rays, high energy charged particles (protons and alpha particles) which create a cascade of secondary particles as they interact with atoms in the atmosphere. This cascade

Fig. 6.8 Combestone Tor, Dartmoor. (Reproduced with permission from Shutterstock, Hazel Mead)

includes a small fraction of hadrons, including neutrons. When one of these particles strikes an atom, it dislodges one or more protons and/or neutrons from that atom. This process is called spallation. The result produces either a new element or an isotope of the original. Now most cosmic interactions occur at the surface or close to it so by using certain cosmogenic radionuclides, you can date how long that surface has been exposed; the more nuclides the older the surface is. The two most frequently measured cosmogenic nuclides are beryllium-10 and aluminum-26. These nuclides are produced when cosmic rays strike oxygen-16 and silicon-28, which as we know are common rock forming minerals. Chlorine-36 nuclides are also used and result from cosmic ray spallation of calcium or potassium.

Back to tors. If we sample the exposed surface of tors then according to the Linton model, they should all have been exposed at a similar time, when the regolith was removed by periglacial processes. In fact, as with most things the story is more complex. There is a strong cluster of ages between 36,000 and 50,000 years ago which corresponds to the middle of the last glacial cycle (termed Devensian in the UK), but there are also much older and younger dates as well. The few older tor ages have no systematic explanation and may indicate inheritance from earlier cycles of bedrock exposure. Since most tors are modest in height and often in advanced stages of disintegration, the general impression is that they emerge and quickly disappear during every Pleistocene climatic downturn.

- Task 6.7: If you are interested in tors or Devon more generally why not check out the work on dating the tors by Gunnell et al. (2013). Alternatively find out where else in the world tors are to be found?

Soils

So, we have rocks, we have agents of weathering which break up those rocks. Weathering is a vital part of the soil formation, and we all depend on soils. This leads us to the question of when does regolith become a soil?

If you are an engineer, regolith and soil are the same thing, since the engineering definition of a soil is anything, you can dig with an excavator. The geographer defines soil as an organically active media, whereas a regolith is biologically inactive. We have regolith on the lifeless moon for example, but not soil. It is the addition of life (and dead life) that makes a soil, the combination of the inorganic products of rock weathering with the organic residue of life, mixed and mediated by life. The principal inputs to a soil, mainly from top down, are humus (decaying leaf litter) and water, with inorganic inputs coming bottom-up from the weathering of rock.

One of the key characteristics of a soil are vertical horizons or layers in which different constituents and processes dominate. We describe these layers from top to bottom as the A-, B- and C-horizons. Before we look at the processes which cause these horizons and give us different soil types I just want to re-emphasis just how important soils are (Haygarth & Ritz, 2009).

To do this we need to introduce the idea of ecosystem services, which are benefits to humans gifted by healthy ecosystems they can include everything from food, clean air, and unpolluted drinking water. These services are often grouped into four broad categories: (1) provisioning, such as the production of food and water; (2) regulating, such as the control of climate and disease; (3) supporting, such as nutrient cycles and oxygen production; and (4) cultural, such as spiritual and recreational benefits. Soils contribute to all of these categories.

Hans Jenny (1899–1992) proposed a conceptual model for looking at soil formations which says it was the sum of:

$$S = f(C, O, R, P, T \ldots)$$

Where C is climate, O is organic matter, R is relief, P is parent material and T is time. Climate controls the movement of water in the soil; down if precipitation exceeds evaporation, and up if evaporation is greater than rainfall. Organic matter or vegetation determines the organic input while the parent material the inorganic input. Relief controls the movement of soil through a point on a slope and the process of soil formation takes time. Vary all or any one of these variables and you get different soils in different locations. Soils are sufficiently homogenous to allow classification and mapping something that started quite late in Britain (1950s) as a consequence of concerns raised during the Second World War over food production. State funding was never maintained, and the UK no longer have a formal soil survey although many other nations do.

Soil horizons develop from the movement of chemical components (leaching) and the physical transport of clays (eluviation) from one horizon or layer to another where they are either re-deposited (illuviation) or ultimately lost to groundwater. Where rainfall exceeds evapotranspiration, this is a top-down process with loss from surface layers and re-deposition at depth. In contrast in arid regions the movement may be bottom-up as the rise of soil moisture is driven by evaporation. Leaching is assisted by combining organic residues with clays and oxides to form complex compounds (chelates) which increase mobility of solutes and of weathering. Figure 6.9 summaries three important soil forming processes. The first is laterization typical in the humid tropics where intense chemical weathering breaks rocks and organic matter down quickly and high rainfall leaches them from the soil to leave deep nutrient depleted soils. The reverse process is calcification where the upward movement of soil waters driven by evaporation leads to the build of salts and bases in upper soil horizons and in extremely arid settings salts may occur at the surface.

Podzolization is common in cool, wet latitudes and involves the eluviation of bases, clays and oxides from surface layers and their re-deposition lower in the soil profile. The process of podzolization is common on the heathlands of Dorset where acidic leaf litter (pines and heathers) combine with sandy, free draining soils to give classic podzols even though Dorset is not typically that cold and wet. You can explore the soils near your home, or those a favourite location using an online tool to be found at: http://www.landis.org.uk/soilscapes/. Soils are classified in different ways depending on the country involved and according to the US Soil Conservation

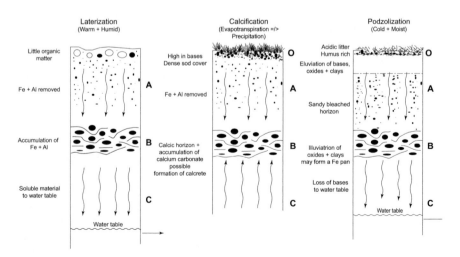

Fig. 6.9 Three common soil forming processes

Table 6.1 Twelve soil orders from the US Soil Conservation Service. Area is defined as ice free land surface

Soil	Characteristic	Area
Oxisols	Hot and humid tropics, high amounts of weathering and leaching leading to laterization and a homogenous soil. Low natural fertility.	8%
Aridisols	Hot, dry deserts. Little alteration of bedrock with surface illuviation of carbonates and evaporites.	12%
Mollisols	Sub-humid and semi-arid grasslands known as chernozems and are rich in organic material. They are some of the most fertile soils – The bread-basket of the prairies and the steppe. Balance between rainfall and evaporation is key maintain lots of salts and nutrients in soil.	7%
Alfisols	Semi-arid to moist areas. Weathering process that leach clay minerals out of surface layer into subsurface. Forest and mixed vegetation and can be productive.	10%
Ultisols	Red-yellow highly weathered soil of subtropical forests	8%
Spodosols	Also known as podzols occur under conifer and cool humid forests, highly leached horizons and lower illuviated horizons. Tend to be acidic and infertile.	4%
Entisols	Recent soils with no evidence of pedogenic development (i.e. profile).	16%
Inceptisols	Embryonic soils in humid and subarctic forests and consequently have a range of characteristics.	17%
Vertisols	Tropical black clays are characterised by expandable clay that leads to development of vertical joints on wetting and drying. They occur in tropics and subtropics.	2%
Gelisols	Soils with permafrost at or near surface.	9%
Histosols	Peats with no permafrost and over 20% organic matter common in UK (bogs and moors).	1%
Andisols	Occur in areas of volcanic activity with lots of volcanic ash providing rich nutrient supply via weathering.	1%

https://www.nrcs.usda.gov/wps/portal/nrcs/site/soils/home/   and   https://en.wikipedia.org/wiki/USDA_soil_taxonomy

Service there are twelve major types (Table 6.1). Note that this is just one of several classification systems which can cause a lot of confusion especially when in the UK we have traditionally used terms like brown earths, stagnogleys and the like!

- Task 6.8: Try to expand on Table 6.1. Can you find a soil profile for each of the soil types? Can you add a column with a fertility score (e.g., arbitrary zero to five stars?). What are some alternative names for the same soils in FAO/UNESCO classification system?

Further Reading[1]

Chen, J., Blume, H. P., & Beyer, L. (2000). Weathering of rocks induced by lichen colonization—A review. *Catena, 39*, 121–146.

Cooke, R. U. (1979). Laboratory simulation of salt weathering processes in arid environments. *Earth Surface Processes, 4*, 347–359.

Eash, N. S., et al. (2015). *Soil science simplified*. Wiley.

Gunnell, Y., et al. (2013). The granite tors of Dartmoor, Southwest England: Rapid and recent emergence revealed by Late Pleistocene cosmogenic apparent exposure ages. *Quaternary Science Reviews, 61*, 62–76.

Haygarth, P. M., & Ritz, K. (2009). The future of soils and land use in the UK: Soil systems for the provision of land-based ecosystem services. *Land Use Policy, 26*, S187–S197.

Natural England. (2020). *Limestone pavement orders.* https://data.gov.uk/dataset/ff9f1088-4b07-4cab-88cf-838a8f421328/limestone-pavement-orders.

Nicholson, D. T., & Nicholson, F. H. (2000). Physical deterioration of sedimentary rocks subjected to experimental freeze–thaw weathering. *Earth Surface Processes and Landforms, 25*, 1295–1307.

Pawlik, Ł., Phillips, J. D., & Šamonil, P. (2016). Roots, rock, and regolith: Biomechanical and biochemical weathering by trees and its impact on hillslopes—A critical literature review. *Earth-Science Reviews, 159*, 142–159.

Smithson, P., Addison, K., & Atkinson, K. (2013). *Fundamentals of the physical environment*. Routledge.

White, R. E. (2013). *Principles and practice of soil science: The soil as a natural resource*. Wiley.

Winchester, S. (2002). *The map that changed the world: A tale of rocks, ruin and redemption*. Penguin UK.

---

[1] This chapter has a real mix of things in it of which soil formation is probably the most important if you are an ecology student. Most textbooks carry a section on weathering and my preferred one Smithson et al. (2013) Fundamentals of the Physical Environment is no exception. It also has some excellent chapters on soil formation with a particularly British perspective. The website of the UK Soil Observatory (http://www.ukso.org/quick-links.html) also contains lots of useful information and relevant links. If you are really keen to know more about soils, there are several dedicated textbooks but bear in mind that these support whole soil science units! White (2013) or Eash et al. (2015) are two such examples.

Chapter 7
Gravity and Slopes

Forces on a Clast(s), or Rock Slope

An object, be it a boulder, pebble, cobble, granule, or sand grain for that matter, are all subject to the same forces as they sit on a slope. Let us set aside finer-grained material for the moment. If our object, let's call it by the geological term clast, or grain if it is small, sits on a horizontal slope then the force acting vertically down is a function of its weight and there will be an equal and opposite force acting upwards. If lateral forces act on the clast, such as the wind or a flow of water, they must overcome both weight of the clast and the contact friction between it and the surface if it is to move. The friction is associated with the contact area between the clast and the ground. The greater the contact area and weight the harder it will be to move the clast. If we now incline the slope on which the clast is sitting both the downward and upward forces become split, between those acting vertically and those acting via the tangent the clast makes with the ground (Fig. 7.1). As the slope approaches the vertical all the force will be acting vertically, and the object will likely be in free-fall.

The individual properties of the clast control the friction, such things as size, shape, and surface texture. Size is easy to understand, the larger the clast, the greater its weight and associated friction. In terms of shape the more spherical the clast is the more likely it is for the clast to roll, similarly a flat, thin stone might slide easily. Rounded or sharp corners will also be a factor along with the surface roughness of the clast. We measure many of these properties by making three axial measurements (a, b, and c axes). A spherical clast will have axial measurements that are similar, and a perfect sphere would give dimensions such that, a-axis equals the b-axis, equals the c-axis. There are lots of different types of shape measures which can be used here, and we normally determine the size of a clast via its b-axis.

If we now increase the numbers of clasts siting on the slope so that we have a granular sediment, they will begin to interact in such a way as too impact on the friction experienced by individual clasts. For example, a smaller clast nestled

© The Author(s), under exclusive license to Springer Nature Switzerland AG 2022
M. R. Bennett, *Our Dynamic Earth: A Primer*,
https://doi.org/10.1007/978-3-030-90351-0_7

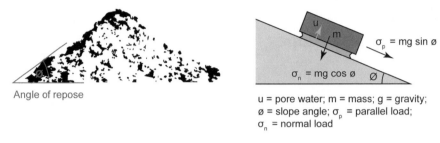

Angle of repose

$\sigma_p = mg \sin \varnothing$

$\sigma_n = mg \cos \varnothing$ $\varnothing$

u = pore water; m = mass; g = gravity;
$\varnothing$ = slope angle; σ_p = parallel load;
σ_n = normal load

Fig. 7.1 Forces on a single or multiple grains (clasts). Looks complex but it is not. The formula simply states that the forces acting on the block become partitioned downslope as the slope angle increases

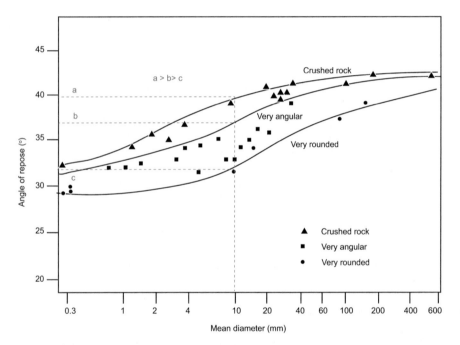

Fig. 7.2 Variation of angle of repose with grain size and roundness. Note how the angle of repose for a 10 mm diameter grain is much greater for angular crushed rock than for a rounded grain. Why? Because the angular grains lock together better. (Simons, D.B., Albertson, M.L., 1963. Uniform water conveyance channels in alluvial material. *Transactions of the American Society of Civil Engineers* 128, 65–107, Figure 12)

between two big ones will be held in place by the larger ones. We refer to this as clast-packing and also via the term sorting. A well sorted granular mass (a sediment really) will have clasts of similar size, while a poorly sorted one will have a range of different sizes. All of these properties determine the internal angle of friction of a sediment. This is basically the angle at which a pile of that sediment will naturally rest at if poured out of a giant jar. The more angular the grains the more they will lock together and the greater the angle of friction will be (Fig. 7.2). A well sorted sediment may be slightly weaker than a poorly sorted one and so on. The internal angle of friction is a measure of the internal friction between the grains in a dry sediment. Dry is the key word here.

When we deal with wet sediment especially finer granular materials, we need to also consider pore water content. That is the water that surrounds the individual grains and fills the gaps (pores) between those grains. Water does two things it acts as a lubricant and can also move particles apart. We talk about pore water pressure, the pressure of the water in the pores. The more water in the sediment the higher the pressure and this can move the grains apart allowing them to move past one another more easily. You can think of pore pressure a bit like a bathtub, the water level or pressure is a function of the rate at which you add the water relative to the rate at which it drains away.

Now everything we have discussed so far works for individual boulders, clasts, or grains in a granular sediment as long as the mean size of the grains is over 63 μm which is equivalent to a grain that is 0.064 mm in diameter. Why this threshold the honest answer is I am not quite sure, in part tradition but also an empirical estimate of when cohesive forces start to have an impact. So, what is cohesion? Well, you experience it when your beer glass sticks to the bar top. The tiny layer of water between glass and top grabs the two due to the surface tension of the water film. You get the same effect between fine-grained particles which pulls them together. To dry and it will not work, to wet and it will not; there is a moisture level that is just right for the size of grains you have. So, cohesion will vary with pore water pressure and will other things being equal weaken as pore water pressure increases. There is another element here and that is the electromagnetic attraction between clay particles. Clay particles have a size less than 0.002 mm and consist of tiny little plate-like fragments. The broken ends of these plates have small electrical charges that can bind together and hold clay particles tightly. So, in sediment with a lot of clay and some water we get both cohesion and electromagnetic attraction.

We can summarise this in a simple equation developed by Charles-Augustin de Coulomb (1736–1806) which states that shear strength of a granular material S is a function of:

$$S = C + (\sigma - \mu)\tan\theta$$

C is a measure of cohesion and in non-cohesive materials is zero, σ is the weight of the material (normal stress) minus the water pressure μ and $\varnothing$ is the angle of internal friction. Cohesion increases with the proportion of clay within a sediment.

What about the cliff or rock surface on which our granular mass sits? Well, this is governed by the weakest link in the chain which could be a host of things. You can take an intact sample of rock and crush it in a machine, and it will appear quite strong varying with the rock type and perhaps how weathered the sample is. While this is a component of the chain it is rarely the weakest link, partings are. A parting is any fracture, joint or bedding surface and is a line of potential weakness. If it is just a hairline fracture the two sides are probably firmly in contact, but as we increase the gap the surface contact decreases and the movement along that fracture become more likely. Add some water, soil or vegetation into that gap and the likelihood of movement increases further. The continuity of partings is also important; are they isolated, or do they all join up so that effectively you can lift out a block? Do the partings dip out of the slope or into the slope? In the second case gravity will help hold the block or layer in place, whereas if the parting dips out of the slope gravity will tend to pull the block or layer out. Collectively these factors contribute to what is referred to as Rock Mass Strength and there are various classifications that can be used to determine this. Some are used by engineers designing cuttings and tunnels, while others have been developed by geomorphologists for more general use.

Whether we are dealing with a rock mass, boulder, granular sediment, or cohesive soil we can conceptualise the forces keeping it in place on a slope as the balance between the forces driving movement versus the forces preventing movement. This balance in the context of slopes is called the Factor of Safety.

$$Factor\ of\ Safety = \frac{Strength\ of\ material}{Applied\ stress}$$

The Factor of Safety is the ratio of slope strength to stress and is usually greater than one. That corresponds to a stress, driving forces if you like, being less than the strength or resisting forces. When the driving forces exceed the strength, we get active failure and the Factor of Safety fall below one. It is a good way of looking at slope failure, what processes will weaken a slope and what forces will increase the stress acting on it? While this is a useful device to look at the causes of failure and engineer would try and calculate, or at least estimate, the stresses and material strengths. We will use this device in the next section when we look at mass movement.

Mass Movement

Any downslope material of rock or soil counts as a mass movement. The term is inclusive of a wider range of process from rock-fall, via soil creep to major landslides and it is an important landscape process by which mountains are eroded. It also has consequences for those living in their path.

In terms of classifying mass movements, we can think in terms of speed of action, type of movement and whether the material involved is dry or wet. These three variables allow us to produce a simple classification (Fig. 7.3). Speed can be deceptive in terms of work done in the landscape. A process that may be of small magnitude but occurs with a high frequency such as soil creep may have a greater impact over a long period of time than a catastrophic failure that while large only occurs once in a blue moon. Small high frequency events are often the most effective tools of landscape change. It is a concept that was developed by Wolman and Miller (1960) in a famous study on rivers and goes under the term of magnitude and frequency (Fig. 7.4).

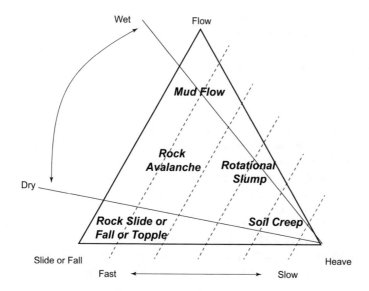

Fig. 7.3 Simple classification of mass movements

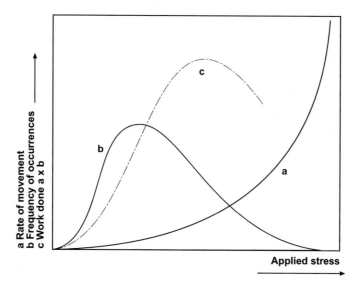

Fig. 7.4 Magnitude and frequency as depicted by Wolman and Miller (1960). The greater the applied stress the more movement can be achieved; small stress events occur with greater frequency than large ones; and so, the maximum geomorphological work is shown by line c. (Wolman, M.G., Miller, J.P., 1960. Magnitude and frequency of forces in geomorphic processes. *Journal of Geology* 68, 54–74, Figure 1)

In terms of process, we can contrast material that slides over a surface or slip plane with that which flows (Fig. 7.3). A saturated mass of mud and water flows as does a ductile clay. We can also think in terms of a heave type process. A grain of soil sitting on a hill side will move outwards perpendicular to the surface slope if driven up by a frost crystal or as the surface expands on wetting. However, gravity will ensure that this particle falls back vertically when the surface dries or the frost crystal melts. In this way the particle moves downslope in a series of steps or is said to have move by creep. Within our broad classification we can contrast dry failures such as big rockslides and falls with wet failures of mud or sand that flow. Irrespective of the type of movement the causes are similar and can be best viewed in terms of the Factor of Safety and those processes which work to weaken the strength of slope materials and those which increase the stress acting on those materials.

The most effective way of weakening slope materials is to weather them either via mechanical or chemical processes. Weathering transforms a rock or soil and thereby reduces its material strength. Since it works from both the surface and along sub-surface fractures it can be highly effective over a number of years depending on how aggressive the weathering regime is. This is in partly likely to be determined by climate and other environmental conditions. For example, the evaporation of saline groundwater can rapidly erode rock surfaces. Variation in pore water pressure can also be effective in varying slope strength both over the long term if there is a progressive increase in water pressure or in cyclic fashion with daily or seasonal variation in rainfall.

The data in Fig. 7.5 is a lovely set of observations. It is a cumulative plot of rainfall at various site in California. A rise in slope gradient signifies more rainfall, while a plateau occurs when there is a pause in rainfall. The dots show mass movement events, in this case mudflows, note how they concentrate on the periods of rainfall where the cumulative rainfall rises. Rainfall saturates the ground and accumulates increasing pore water pressures which in turn reduces the strength of the slope materials as the grains are pushed apart. There are lots of examples like this and the link between rainfall, storm events and episodes of mass movement is clear just search on YouTube for landslide videos and invariably most of them are shot in the rain! It is not always just about rain; however, damaged drains can lead to slope failure to. In areas prone to mass movement maintaining good drainage and ensuring that natural soak-a-ways are avoided is also key. A damaged or blocked drain can be just as effective at building pore water pressure as a more natural process especially if you wash your car frequently.

On the other side of the equation, we have stress. Perhaps the simplest way of increasing the stress on a slope is to increase its weight or load. Small acorns grow into big trees and a large oak tree can weight up to 14,385 kg which in imperial tons is around 14. The total branch-wood actually weighs more than the tree's trunk, accounting on average for 58% of the total weight. Clearly the acorn to fully grown tree has the potential to add a substantial load. The accumulation of rock debris from upslope, or direct human intervention such as buildings are also ways of increasing the load. Removal of lateral support from the base of a slope is also effective in increasing the slope angle and the force acting downwards. Earthquakes,

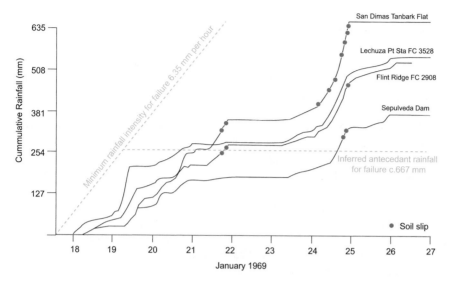

Fig. 7.5 Data on cumulative rainfall near the Santa Monica and San Gabriel mountains in California. Note the occurrence of soils slips with increases in rainfall. (Campbell, R.H., 1975. Soil slips, debris flows, and rainstorms in the Santa Monica Mountains and vicinity, southern California (Vol. 851). US Government Printing Office. Figure 9)

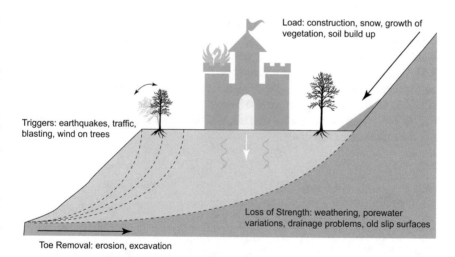

Fig. 7.6 Summary of the causes of mass movement

traffic, or tree vibration may also provide transitory and potentially episodic increases in stress and are effective in triggering a failure. Figure 7.6 provides a summary of some of the causes of failure.

- Task 7.1: YouTube is a great source of videos on landslides and rock falls, take a look and find some examples to supplement your notes.

In terms of types of mass movement, we can make a general distinction between hard-rock failures and soft-rock failures. Hard-rock failures involves failure along joints and partings. The geometry and size of these failures is largely determined by the properties of the rock mass and can range from individual block falls to large topples and slides. Figure 7.7 illustrates how the pattern of joints determines the geometry of the failure. Joints or bedding surfaces dipping out of the slope are more prone to failure than ones dipping into the slope.

Soft-rock failures involve the deformation and/or failure of the rock or sediment itself. This may occur as a flow, in which the entire body of rock/sediment is mobilised or as a slide along a particular failure plane. If the whole body of sediment/rock is in motion we get something called a debris flow, if failure is more localised on a slip plane, then we might get a landslide or rotational slip. The slip plane may be located along a bedding plane where there is a change in permeability or strength for example or in clays it may be located along a line of maximum stress. In clays these slip planes tend to be curved and lead to progressive rotation of blocks. Figure 7.8 shows a series of modelled failures which illustrates the progressive nature of some rotational mass movements.

- Task 7.2: There are some fantastic examples of rotational and complex failures along the Jurassic Coast close to Lyme Regis in the UK. What can you find out about these mass movements? Alternatively research a major mass movement in another country.

There is one type of mass movement which is important but far from spectacular and that is soil creep. It is a little and often type of process which involves the upward movement of a soil grain perpendicular to the slope surface and its subsequently vertical fall. The steeper the slope the greater the downslope step. The 'heave' component may be due to wetting and drying, frost action and the growth of ice crystals below the soil grain, or due to root expansion. The impact of soil creep can often be seen in the verticality of trees, with bent or bowed trees being common under the influence of soil creep.

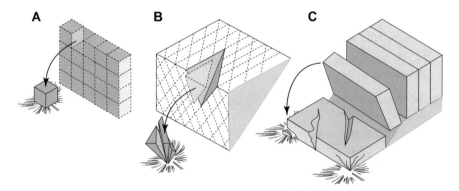

Fig. 7.7 Types of hard-rock failure determined by the joint pattern within the rock mass

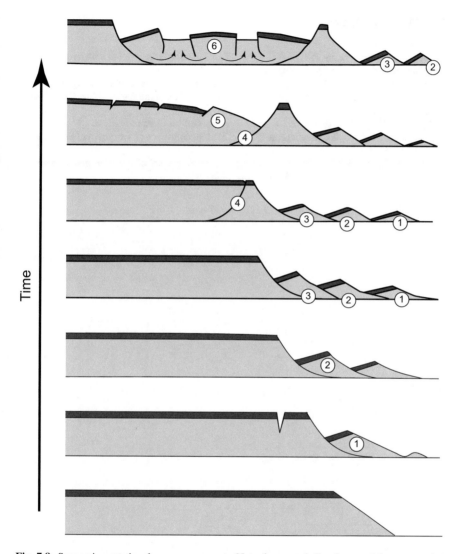

Fig. 7.8 Successive rotational mass movements. Note the curved slip plane and the progressive failure pattern A to G. (Seed, H.B., Wilson, S.D., 1967. The Turnagain heights landslide, Anchorage, Alaska. *Journal of the Soil Mechanics and Foundations Division* 93, 325–353. Figure 22)

- Task 7.3: There is a great review paper by Pawlikab and Šamonilc (2018) which provides a definitive survey of the literature and theory behind soil creep. It is worth looking at and you may find the sections on impact on trees interesting.

In summary mass movements come in all shapes and sizes but have one thing in common which is that failure only occurs where the stress acting exceeds the materials strength.

Further Reading[1]

Highland, L., & Bobrowsky, P. T. (2008). *The landslide handbook: A guide to understanding landslides* (p. 129). US Geological Survey.

Pawlik, Ł., & Šamonil, P. (2018). Soil creep: The driving factors, evidence and significance for biogeomorphic and pedogenic domains and systems–A critical literature review. *Earth Science Reviews, 178*, 257–278.

Selby, M. J. (1993). *Hillslope materials and processes*. Oxford University Press.

Smithson, P., Addison, K., & Atkinson, K. (2013). *Fundamentals of the physical environment*. Routledge.

Wolman, M. G., & Miller, J. P. (1960). Magnitude and frequency of forces in geomorphic processes. *Journal of Geology, 68*, 54–74.

---

[1] Most textbooks carry a section on mass movement and my preferred one Smithson et al. (2013) *Fundamentals of the Physical Environment* is no exception. If you are interest in slope process there is a more detailed book written by M.J. Selby (1993) *Hillslope Materials and Processes*, published by Oxford University Press but it is rather advanced in content. The Landslide Handbook published by the USGS and written by Highland and Bobrowsky (2008) has some great illustrations in it. Just note that they are using the term landslide incorrectly as defined here, they mean mass movements.

Chapter 8
Flowing Water

When it rains, it rains! And rain has a huge capacity for geomorphological work as does ice as we will see in the next chapter. Perhaps the best place to start is with the hydrological cycle. At its simplest evaporation of water from marine or inland water bodies along with evapotranspiration from plants creates water vapour. It is worth noting that this involves environmental cooling since latent energy is used to vibrate the water molecules and effectively liberate them. This water vapour is concentrated and transported before it condenses (releasing latent heat) to form water droplets, clouds and ultimately precipitation in the form or snow, hail, or rain. This water returns to the sea, inland water bodies and to the soil to complete the cycle. We can break this down into four main components, the first encompassing everything from evaporation through to precipitation, the second being the action of water of hill-slopes, the third is surface hydrology and finally we have channel flow. In this chapter we will consider each in turn.

All About Rain

Plants transpire and this helps the flow of nutrient laden water from the roots as well as helping to regulate the osmotic pressure in some cells. Water bodies of all sizes are also subject to evaporation and collectively evapotranspiration is controlled by temperature, wind velocity and local humidity. The latter is a measure of the moisture content a body of air can hold at a given temperature and pressure. Most evaporation occurs at the earth-atmosphere boundary. In order to understand what happens next, we need to think of this boundary layer as a potential envelope of air (Fig. 8.1). What happens to this envelope of air is a function of atmospheric stability.

Cold air is denser than warm air; for a unit number of air-molecules they will occupy less volume when cold, since they are less energetic and therefore closer.

© The Author(s), under exclusive license to Springer Nature Switzerland AG 2022   111
M. R. Bennett, *Our Dynamic Earth: A Primer*,
https://doi.org/10.1007/978-3-030-90351-0_8

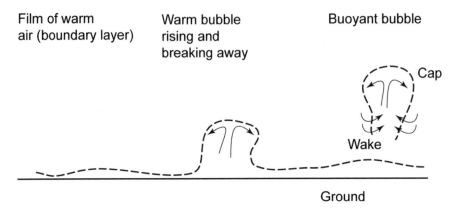

Fig. 8.1 Boundary layer and development of a warm air cell or bubble

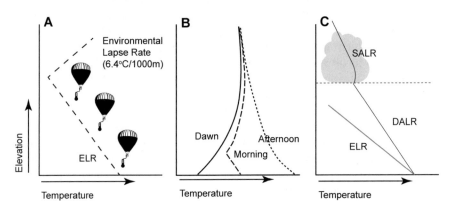

Fig. 8.2 Atmospheric stability. (a) Changing temperature with elevation measured from the balloon gives us the environmental lapse rate. (b) The environmental lapse rate may change through the day as the ground warms up and heats the boundary layer. (c) Dry and saturated adiabatic lapse rate. If they are greater than the ELR at a given elevation air will rise

Now if our envelope of air is warmer than the surrounding air, being heated by contact with the ground it will be less dense than the surrounding air. It should therefore rise upwards until its temperature is the same as the surrounding, or environmental, temperature. On average air temperature falls by about 6.4 °C for every 1000 m of rise something that is referred to as the Environmental Lapse Rate (Fig. 8.2).

This is very much an *average* since the precise temperature gradient may vary during the day as shown in Fig. 8.2b. As we raise our parcel of air it will cool due to the decrease in pressure, the molecules are further apart and put crudely bump into one another less frequently which generates less heat. This rate of cooling assuming that our air is water free occurs at 9.8 °C per 1000 m and is referred to as the Dry Adiabatic Lapse Rate. Adiabatic simply refers to a system that is closed. Now if our parcel of air is warmer than the surrounding air it will continue to rise and will only stop when the air beyond it is the same temperature or colder. This is the essence that determines whether a body of air is either stable, or unstable. The latter rises and cools.

Most air is not completely dry and will contain water vapour, now as the air parcel cools condensation will occur, which due to latent heat gives out heat and warms the air parcel. The more water vapours the more its condensation which will offset the cooling and keep it rising. We refer to this rate of cooling as the Saturated Adiabatic Lapse Rate and it has no standard value since is dependent on the water vapour present in a specific air parcel. The relationship between these various lapse rates gives us different states of atmospheric stability; the atmosphere is unstable if both the SALR and DALR exceed the ELR (Fig. 8.2c).

Atmospheric instability is what drives intense convective summer storms and is one way we can create clouds and ultimately rain. Contact with a cold surface is another way. Warm, saturated coastal air moving on land will often condense to form fog or low-lying cloud banks as it contacts the cooler land. Remember that land naturally heats and cools faster than water; it has less thermal inertia. Another way is by mixing parcels of air with different temperatures and therefore vapour pressures.

Once condensation is underway, we have the first steps in the formation of clouds. Clouds are simply collections of water droplets. There is one additional wrinkle here, that is condensation is favoured by the presence of condensation nuclei which are tiny fragments of dust that help initiate the transition of vapour to liquid. Deserts whether cold are warm have an important role to play in supplying condensation nuclei as do volcanoes and some human pollutants. We saw in Chap. 4 how certain types of plankton in the oceans create dimethyl sulphide which is another source of nuclei. Without nuclei our parcel of air may become supersaturated with either water or ice. Freezing is also favoured by the presence of nuclei or more commonly other ice crystals.

The size of a droplet of water or ice in a cloud varies over time and will progressively grow until it can no longer be held aloft. The rate at which something falls due to gravity is determined by Stokes Law. In 1851, George Gabriel Stokes described the force of viscosity (Frictional Force F_d) on a small sphere moving through a viscous fluid in the following way:

$$F_d = 6\pi\mu Rv$$

Where, μ is the dynamic viscosity a measure put crudely of the 'stickiness' of the fluid or in this case the air. R is the radius of the spherical object which in our case is the water droplet. Finally, v is the flow velocity relative to the object which in our case is the updraft. We can simplify this, such:

$$\frac{1}{v} \propto R$$

As the radius of our droplet (R) increases the updraft (v) must increase to prevent the particle from settling. The stronger the updraft, or instability in the air mass, the larger the droplets can become. If, however uplift or instability is weak then the droplets can not grow much before they begin to fall out. It follows that atmospheric instability is associated with large, heavy rainfall, which by definition will be short-lived since the moisture will be used up quickly. In contrast, weak uplift favours small drops and drizzle which lasts much longer since it takes longer to empty the cloud putting it crudely.

The strength of atmospheric instability is also manifest in the shape and form of clouds. We owe cloud nomenclature to a chap called Luke Howard (1772–1864) who was a British Chemist and an amateur Meteorologist with a broad interest in science. Hamblyn (2001) provides an interesting biography of the man. In 1802 Howard presented a talk with the basic three-fold classification of clouds still in use today: cirrus, stratus, and cumulus. While developed on purely descriptive grounds it is consistent with cloud physics which is why it has endured. He wrote his ideas up in an *Essay on the Modification of Clouds*, which was published in 1803 which contains some great illustrations (Fig. 8.3). Today the best place to go for cloud images and descriptions is the International Cloud Atlas which you can find online.

F.M Williams lith.   X.b.l.f.rbl.sr. .t.s

CUMULOSTRATUS FORMING, FINE WEATHER CIRRI ABOVE

Fig. 8.3 An illustration from Luke Howards (1803) Essay on the Modification of Clouds. (Source: www.commons.wikimedia.org/w/index.php?curid=88440592)

Cirriform clouds are fine upward curving streamers of precipitation (mainly ice) in the upper atmosphere and form in slowly rising air. They often forewarn of the arrival of a weather system. Stratiform clouds are the result of widespread cooling and a stable atmosphere and are formed either by rising air forced over a mountain range for example or by contact with cold ground. They have diffuse outlines and wide horizontal extent and are poorly developed vertically. They are generally associated with fine persistent rain or drizzle. The lack of a strong vertical updraft means that the droplets remain small.

The development of rain drops in stratiform clouds owes something to a theoretical process described by the Swedish Meteorologist Tor Bergeron, building on work of Alfred Wegner (yes, the continental drift guy) and later expanded by Finlayson. It is sometimes therefore referred to as the Wegner-Bergeron-Findeisen process. It is cold in the upper troposphere typically $-50\,°C$, now in the absence of condensation nuclei pure water will not spontaneously freeze until about $-40\,°C$ unless it hits something like an aircraft. Yes, I know you all thought that water froze at $0\,°C$, but not when suspended in air; we call it supercooled water. If this supercooled water comes into contact with a crystalline solid (freezing nuclei) it will freeze; silver oxide is an example and most freezing nuclei become active around $-10\,°C$. So, at temperatures between 0 and $-10\,°C$ most clouds are made up of supercooled droplets. Between -10 and $-20\,°C$ there is a mix of ice and water droplets and below $-20\,°C$ just ice crystals. The key here is the air with a mix of water and ice. In Stratus clouds over lain by cirrus this may also occur by ice crystals falling into the stratiform clouds from above. Now at low temperatures air may be saturated with respect to supercooled water but oversaturated with respect to ice, basically ice is more stable than liquid water. So, at $-20\,°C$ we can have 100% humidity with respect to water but 121% with respect to ice. This oversaturation with respect to ice causes the ice crystals to grow rapidly at the expense of the water droplets. These large ice crystals fall and turn back to liquid water as things warm up lower in the troposphere. To put it simply at these critical temperatures ice crystals, grow large rapidly and much faster than a normal water droplet would given the updraft. Cirrus clouds which are ice rich above stratocumulus is an ideal combination for this process.

In contrast, Cumuliform clouds have strong vertical development and form in an unstable atmosphere where there is often visible evidence of rapid uplift due to convection. These clouds have a heaped appearance with domed tops and clear outlines. They are associated with storms, short bursts of heavy rain, with large drops, hail, and thunder. A convective storm has a clear life cycle starting with towering cumulus clouds which rise to the edge of the troposphere where they spread out as cumulonimbus clouds which are often anvil shaped. The air cannot rise above the Troposphere due to a temperature inversion, so cloud development become height limited. They are associated with intense rainfall and drops of ice rise up in the heart of the convective cell before falling forward, to be swept up again. Each cycle of uplift and fall gives a layer of ice to hail stones whose size is only limited by the strength of the convective uplift.

Convective storms are driven by convection; rapid heating of warm moist air in the boundary layer between the ground and the atmosphere leading to intense atmospheric instability. The convergence of air masses, along the line of Inter Tropical Convergence is an enhanced form of this and the Hadley Cell we discussed in Chap. 3 is driven by convection. The monsoonal systems of Indian and Africa are another example where convection combines with near-surface convergence to drive atmospheric instability and uplift. We get summer convective storms occasional in Britain but not on the same scale as inland continental areas far from the moderating effects of the sea. In most mid-latitude areas such as Britain, Europe, or along the northern Pacific Coast of North America rain is driven by cyclones or as they are more commonly called depressions (not in mood, but in pressure), or low-pressure systems.

We saw in Chap. 3 how low- and high-pressure systems were spurned by convergence or divergence in the westerly jet stream (Rossby Waves). Convergence of the jet stream creates high pressure and descending air which spins clockwise due to the impact of Coriolis Force in the northern hemisphere (anticlockwise in the southern hemisphere) and creates a near-surface rotating cell of warm air and high pressure. In contrast divergence, or widening of the jet stream, leads to low pressure and air to put crudely is sucked upwards. This creates a cell of near-surface air rotating around a low-pressure centre. Rotation is anticlockwise in the northern hemisphere and clockwise in the southern hemisphere. In the case of both cyclones and anticyclones once they are set spinning, they then move freely with the surface westerlies gradually decreasing in spin before dying out. Exactly the same as happens when you set a spinning top in motion and then watch it move over the surface before it finally stops. In the case of the Atlantic pressure systems form on the west coast and then move across the ocean making landfall on the western seaboard of Europe. If you are confused by the differences in spin cyclonic spin in the two hemispheres there is a bit of weather-lore that might help. Buy's Ballots Law states that if you stand with your back to the wind the low pressure will lie to your left in the northern hemisphere and to your right in the southern hemisphere.

Cyclones and anticyclones are an important part of secondary global circulation and are like giant spoons mixing air masses, warm air from the south with cold air from the north. It is the mixing of different air masses that creates rainfall within a cyclone. Meteorologists call the boundary between different air masses a weather front. On a typical cyclone there are three types of fronts. A cold front occurs where a cold air mass, which is naturally denser, moves below a warmer air mass. The front is often steep, and the warm air is forced upwards rapidly causing condensation and cumulonimbus or nimbostratus clouds to form. These clouds give heavy but usually short periods of rain occasional associated with hail and even thunder. The passage of this cold front is usually marked by a rise in temperature at the ground as one moves from cold to warmer air. The second type of front, a warm front, occurs where warm air moves more gently over cold air. Its passage is again heralded by a slight rise in air temperature but also by a classic cloud sequence cirrus, cirrusstratus, altostratus and finally nimbostratus. Rain fall is more prolonged, finer, and often with a large component of drizzle. Where the warm air mass is completely lifted from the ground by cold air, effectively by the warm and cold fronts meeting, you get an occluded front. The steepness of the fronts and the rainfall within a cyclone is a function of the intensity of the low pressure and the contrast in terms of moisture and temperature of the two air masses being mixed. The more intense the low-pressure the stronger the winds and the greater the contrast in air masses being mixed the more intense the weather system and rainfall will be.

The routing of the Rossby Waves or westerly jet stream has a profound effect on the local meteorology of the eastern Atlantic and Europe. This is because it determines the spawning of the cyclones and anticyclones and their path across the Atlantic (Fig. 8.4). A significant part of the annual variability in both the summer and winter seasons in a country such as Britain is determined by the jet stream, and it is one of the facts that makes Britain such a lovely place to live unless you like predictable weather!

One final point here and that is snow. We have talked a lot about rain but there are other types of precipitation and we have seen how rain often forms as ice crystals. The type of precipitation, frozen or not, one experiences on the ground largely depends on the vertical temperature profile. If that profile never gets above freezing or not till very close to the ground, then your precipitation will invariable be frozen. Since that profile is in part determined by the season, summer snowfall is an exception not the norm.

One of the reasons why meteorologist tend to talk about low-pressure systems or depressions is because the term cyclone is also used for a type of intense tropical low-pressure system which also goes under the name of hurricane and typhoon. While and intense low-pressure system it is not formed in the same way as mid-latitude cyclones although it may become one in its death throws. Hurricanes, as they are called in the Caribbean and Atlantic only occur in warm tropical waters generally having surface water temperatures over 27 °C which are normally found 5° N and S of the equator. They tend to be a feature of the late summer and early autumn and their frequency is variable. Typically wind speed exceeds about 250 km per hour around the eye of the storm. On average, over the past 50 years tropical

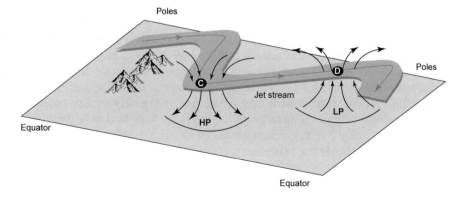

Fig. 8.4 Role of Rossby Waves in generating cyclones and anticyclones in the mid-latitudes

cyclones have killed 10,000 people and incurred damages of $15 billion annually around the globe. However, this has to been seen in the light of increased human vulnerability; over this interval, the population exposed to tropical cyclone hazards tripled, owing to migration into coastal areas. So, having looked briefly at the first part of the hydrological cycle we now need to consider what happens to the water once it falls on land.

- Task 8.1: Pick a couple of hurricanes in the last twenty years and find out some facts, figures and impacts about them.

Water on Hillslopes

When it rains on a hillslope three things can happen to that water. It can be inter-cepted by a woolly sheep (or equivalent), or caught on a stem or leaf either way it is lost to the system, and we need not consider it further here. The alternatives are that it hits the ground and soaks in, or it hits the ground and flows over it usually in a downslope direction. The choice between infiltration and runoff to use the correct terms is a function of the permeability of the soil/rock, the antecedent soil mois-ture conditions, and crucially the rainfall intensity. Most rocks are covered by soil or regolith so there is some infiltration capacity even if the rock is completely impermeable to water. Antecedent simply means previous or past, so if it has rained a lot the soil/rock may be damp, and any water storage capacity may be full just like a wet sponge. The key is rainfall intensity however because you can get runoff even when the soil is bone dry. Think of a sieve with fine mesh; if the water fall onto the surface faster than it can drain through the holes then it will pond and run of the surface. The same is true of soil. If you add rainfall to quickly then it cannot soak in fast enough. In this case the rainfall intensity which is simply a volume of water in a unit of time exceeds the infiltration capacity. This is why in arid deserts you often have flash floods; the intense convective rainfall simply cannot infiltrate fast enough into the ground. Where we get runoff in this way (intensity > infiltration) we call it Horton or Hortonian runoff.

Now we can also get runoff simple because the ground is saturated with mois-ture. We call this saturated overland flow. It is common at the bottom of slopes or along stream banks where water collects from upslope adding to the saturated nature of the soil and/or rock. Overland flow of whatever type can erode, transport and deposit sediment on a hillslope, it is an agent of change. It usually starts as sheet flow, a thin layer of water flowing over the surface, but quickly becomes concen-trated into rivulets (rills) and in turn into larger gullies. Rills tend to be ephemeral while gullies usually persist between rainfall events. The transition from sheets to rills and gullies is largely due to random processes; bushes, large rocks, boulders, or topographic undulations all tend to focus the water initially into a flow which then persist.

Erosion is achieved by a range of processes and is countered by vegetation and an absence of a slope gradient. The steeper a slope and the longer the slope is uninterrupted the stronger the water flow and the greater the capacity for erosion. Also, the more vegetated the slope is in terms of ground cover the less erosion takes place. Roots are fantastic at binding soil and stems disrupt and baffle water flow. Semi-arid regions erode fastest because there is a lot of bare soil, and enough rain. Deserts have more bare soil, but it rains to infrequently for much erosion to occur.

Sediment entrainment (erosion + transport) involves fluid drag and also lift. The lift comes from the flow of water around the curved surfaces of a grain, in a similar way you get lift around a hydrofoil or aeroplane wing. Drag is a function of water velocity and turbulence in the water column. Entrainment by fluid drag will increase with water velocity and smaller grains should be entrained more easily than larger ones. These properties can be summarised theoretically in what is known as the Hjulström Curve, named after Filip Hjulström (1902–1982). The curve plots water velocity against grain-size and defines three regions one for erosion and others for transport and deposition (Fig. 8.5).

If we look first at the erosion threshold on the Hjulström Curve we see that as one might predict that erosion requires stronger current velocities as we increase grain size of the bed beyond 0.1 mm. The odd bit is that before that current velocity declines with increasing grains size. This is because of the effects of cohesion and the more clay you have in the sediment the more cohesive it tends to be. So initially as we increase the grain size the threshold for erosion falls, before rising again after 0.1 mm. Cohesion does not come into play for coarse sediment.

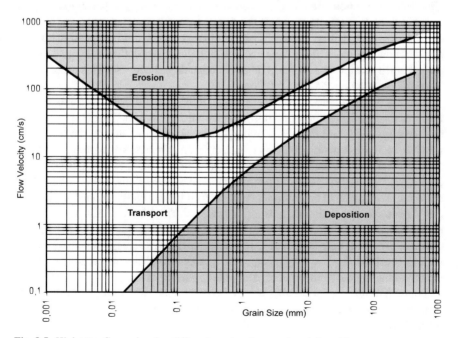

Fig. 8.5 Hjulström Curve showing different erosion, transport, and deposition zones

The relationship between deposition and velocity is simpler, as current velocity falls, we get deposition for a given grain size. Note however that it is very hard to deposit small grains. Anything below about 0.01 mm is almost impossible to deposit. The fact that we do get such fine-grained deposition owes itself to the process of flocculation which binds small grains together to make larger ones. This is important for deposition on mud flats for example. The other thing to note in the Hjulström Curve is that for a given grain size the erosion and deposition lines are separated. In general, it takes more flow energy to erode a grain than it does to deposit it. The reason for this is that is erosional inertia. Grains tend to stick or pack together and are difficult to move initially, once they are moving however, they have momentum, and it is easier for them to keep moving. This curve is one of the fundamental diagrams in physical geography and while it cannot be used effectively to predict erosion in a given channel precisely the shape of the curves conveys a lot about the fundamental physics of sediment entrainment. The curve is also applicable to erosion and deposition within a channel and in the oceans.

One final point before we move. The processes of rill and sheet erosion have been subject to lots of experimentation. It is relatively easy to set up a lab experiment with a sprinkler and large tray of soil, or to set up an experimental plot outdoors. One interesting study tried to quantify the relative importance of rill and sheet erosion using nuclear fallout (Whiting et al., 2001). Ever since our species exploded and tested nuclear bombs and had nuclear disasters of various sorts there has been fallout. Different radionuclides exist at different depths in a soil profile. As a result, if erosion taps a particular depth of soil, then it will give a unique radionuclide signature. Another depth gives another signature. So, on a special agricultural test plot these researchers captured the runoff from a particular storm and worked out the isotope signature present. They compared this to the isotope signature by depth obtained prior to the erosion event. Using this method, they were able to estimate that rill erosion was 29 times more effective than sheetwash. This makes sense since rill erosion is a concentrated flow.

There is one more process that operates on hillslopes and that is rain splash. This involves the physical impact of a rain drop with the ground and the displacement of surface sediment grains thereby. The bigger the rain drop the greater the fall velocity and the more kinetic energy is released when it contacts the ground. Intense convective storms are ideal for generating high impact rainfall events. On a horizontal surface the drop displaces sediment grains in all directions, but on an inclined surface gravity kicks in and sediment is displaced further on the downslope side. In this way downslope sediment transport is achieved. Yes, I know that it is a low magnitude process, one raindrop at a time, but if it rains enough the process can be effective provided there is a lot of bare soil/sediment. Rain splash can also enhance Horton overland flow by displacing small grains into sediment pores and cracks thereby reducing the infiltration capacity of the surface layer.

Figure 8.6 shows an interesting piece of experimental data from a controlled field plot which shows how rain splash and sheet wash combine to be effective agents of slope change.

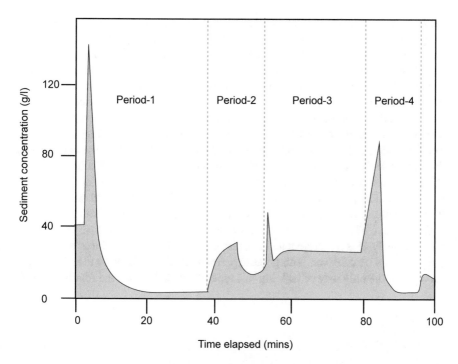

Fig. 8.6 An experimental plot of sediment erosion which is measured as a concentration in the water flowing of the plot versus time. In Period 1 there is overland flow only; Period 2 overland plus rain splash; Period 3 rain splash only; Period 4 overland flow only; and Period 5 overland flow plus rain splash. (Ellison, W.D., 1945. Some effects of raindrops and surface-flow on soil erosion and infiltration. *Eos, Transactions American Geophysical Union 26,* 415–429. Figure 11)

In the first period erosion is rapid but declines over time not because of any change in the sheet wash velocity, but because the available, loose surface grains are all removed (Fig. 8.6). Put simply all the easily moved material is moved and supply begins to lessen up. If we introduce a period of rain splash (Period 2) then we dislodge and loosen new grains and erosion increases. In Period 3 we only have rain splash operating and initially it is effective but settles down to a steady rate quite quickly. In Period 4 overland flow returns and sweeps up all the loosened material and we again have a peak in erosion. Although this is an artificial plot in which we can turn rain and sheet wash on and off it shows how the two processes can work together to achieve effective slope erosion. It also shows how good data can be obtained from some simple experimental plots.

There is one other bit of research that I would like to draw your attention to. It is a well designed study to explore the variation in slope morphology with climate and was conducted by Toy (1977). The reason that I draw attention to it is that it illustrates how one can design a landscape experiment that simplifies the variables involved so the focus is on the question in hand. There are many factors which controls slope morphology, geology, climate, local climate including slope aspect, land use and so on. Toy (1977) wanted to isolate climate. He selected 29 sites along two traverses, one set of sites were located close to the 37th parallel which reveals variation in summer rainfall. The other set of sites were located along the 105th Meridian to examine temperature. All the chosen sites were located on sub-horizontal marine shales and with a southerly aspect and were within 5 miles of a weather station with at least 21 years of record. The slopes were surveyed in the field and data on length and curvature extracted and regressed against the weather data. Crudely the slopes had three components a convex upper section, a rectilinear middle section and concave lower section. Approximately 59% of the curvature of the convex section could be explained by climate variables and 43% of the variation in the steepness of the rectilinear section could be explained by climate. Toy (1977) was able to derive a simple model to explain this (Fig. 8.7). Humid areas seem to give straighter and less curved profiles and are the result mainly of soil creep while at the arid end of the spectrum rain splash and sheet wash give a more curved profile. The analysis is not perfect and has attracted some detractors since, but it shows you what can be achieved with good experimental design. Back in the 1970s field survey was the only way of tackling this type of problem hence the relatively small

Fig. 8.7 Sample locations and conceptual model of slope morphology with climate. (Toy, T.J., 1977. Hillslope form and climate. Geological Society of America Bulletin, 88(1), pp. 16–22)

sample. Today you can call up accurate digital elevation data of almost any slope and it would be interesting to see given the availability of such 'big data' whether you would get the same result.

- Task 8.2: Let us assume that you have been asked to design research to address the question 'how does hillslope form change with climate'. How would you approach this using the tools available today? Jot down some ideas and data sources.

Surface Hydrology

In theory the surface of the Earth is a bit like a series of nested funnels, with each funnel directing and concentrating water as part of a bigger funnel. Think of a tiny stream in an upland made of impermeable rock; this is our first funnel. Rain falls within the catchment area of our stream, and it leaves via the stream which is equivalent to the spout. A simple input, storage, surface loss, and output system. Now let us assume that this stream is one of many tributaries feeding a larger river. Each tributary has its own catchment area (drainage basin) but together they sit within the larger basin of the big river. And so, the nested system grows with each new branch; the drainage basin of each stream and river sits within a larger one until you get to the highest order stream (i.e., with most tributaries) and it flows into the sea or into an inland lake. The boundary between one drainage basin and the next is called the interfluve.

At a continental scale we can look at where the continental divide lies; that is the line along which water flows either to one ocean or another. Figure 8.8 is a map of these mega-scaled divides, note that inland draining basins are called endorheic basins. Interestingly 48.7% of all freshwater drains into the Atlantic and just 13% into the Pacific. This is one of the reasons why the thermohaline circulation operates the way it does (Chap. 4) and is a function of the distribution of mountains around the globe and therefore ultimately plate tectonics.

Let us go back to our simple river system with a few modest tributaries. We can represent this via a wireframe in which we have the input from rainfall, and the output at the most downstream river. If our drainage basin sits on permeable rock, then we will also have a component of groundwater either stored within the basin or potentially transferred into another depending on the geometry of the water table.

A small proportion of the rainfall will fall directly into the river or one of its tributaries. Most however will fall on the slope and either flow to the channel as Horton or saturated overland flow. Some may be stored locally within the soil or taken up by roots to be transpired by plant. Another component may be caught on the leaves, stems or branches of a plant and evaporated directly back. Evaporation may also occur directly from puddles, ponds, lakes, streams, or rivers. Some water will also find its way into the bedrock assuming it is permeable and may be stored as part of the water table. There are lots of routes, some more direct than others, by

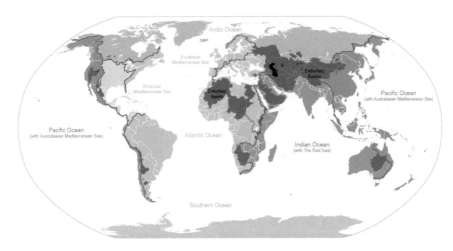

Fig. 8.8 Continental drainage divides. (Source Citynoise at English Wikipedia – Transferred from en.wikipedia to Commons. Public Domain, https://commons.wikimedia.org/w/index.php?curid=3154822)

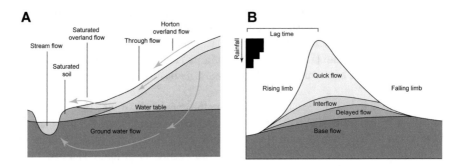

Fig. 8.9 Components of the hydrograph

which water may find itself into the stream and river. Not all of it will make it. These collective properties of the drainage basin are defined by its geology, geomorphology, channel networks, soils, and vegetation. As humans we can impact on this; for example, a large carpark may accelerate the flow of water into a stream, whereas planting trees may slow this down. Landuse can have a profound effect on surface hydrology.

- Task 8.3: Try to draw out the wireframe that represents the various component parts of the drainage basin. Think of it as a flow diagram involving the movement of water. Attempt this without the internet first and then find one of the many wireframes available and compare your version with a published one.

We measure this collectively via the hydrograph (Fig. 8.9). This is a graph that portrays discharge (Q) against time with the rainfall received by a catchment plotted as well. Discharge is a unit volume of water usually in cubic metres in a unit of time. If you know the cross-sectional area of channel, or your gauging station, then it can be measured simply via changes in water depth. We can break the hydrograph into a series of elements. The total envelope of a flood event that follows a discrete episode of rain has a rising limb, peak and falling limb. The area under this line is the total volume of water involved. We can break this down into dry flow (base flow) which is that supplied by groundwater and is present when there is no rain. Above this base there is water that get to the gauging point quickly (quick flow). Maybe it falls directly into the stream. Then we have interflow which is delayed by the processes of surface runoff, and we have delayed flow which is actively held up by something along the line, may be a large puddle or in a pond. The time between the peak and the rainfall event is known as the lag time and is a function of the drainage basin characteristics.

If we take an arid drainage basin for a moment and an intense downpour then the lag time may be short, which is why we get flash floods. The rain is so intense that input to the slope exceeds the infiltration capacity of the soil even though it is dry. Rapid Horton runoff results and the water ends up in the channel quickly, rising the stage (water depth). In contrast a drainage basin with lots of trees and vegetation may have a longer lag time. Think about all those branches and leaves which intercept the rainwater, and we also have lots of roots and plant uptake. One of the best forms of flood protection is afforestation but try telling that to an irritate resident who has just been flooded: "we are going to plant lots of trees, and in about 30 years the flood risk to your house will be gone."

One of the key concepts in studying hydrographs is the idea of the 'unit hydrograph'. If we do not change the land use in the drainage basin then we do not change the 'plumping' linking slopes to stream. As such the shape of the hydrograph should always be the same; the size of the flood, effectively the area under the graph, may change with rainfall but the overall shape will not. This is a really useful tool in flood management. While one could build river defences to stop flood waters, the alternative is to consider how you can change the hydrograph shape. Basically, how can you attenuate, flatten the curve, so that same amount of water is discharged over a longer period of time? That way the river level can be kept below a certain depth. Trees are one solution, not always maintaining drains are another, but perhaps the best is to allow flooding of low-lying agricultural land upstream of a town so that water is stored and release over a longer time. This is beyond the scope here but managing the hydrograph is key.

Channel Flow

We now have our water in a channel, be it a stream, or river. We can divide channels into three broad types: rock floored channel; mixed channels with alternating reaches of rock and sediment; and alluvial channels. A river may contain reaches of all three types at different locations. Alluvial channels have boundaries which are completely made of sediment, typically they occur in the lower reaches of a river. They are well studied, and their dynamics are consequently well understood and as we will see there is a certain elegance about the link between discharge and channel morphology. In contrast rock channels are less well understood being geomorphologically less responsive (i.e., it takes time for things to change) and are more hostile to instruments. Delicate scientific instruments do not take kindly to being battered by rocks rolling along the bed of a stream. We are going to look briefly at the two endmembers, rock-channels, and alluvial channels, although we will probably talk more about the latter and that is where we should therefore start.

Alluvial Channels

In theory a channel within sediment is easily modified by erosion and deposition. As such it should be in approximate equilibrium at least with the discharge or water level that does most work. Let us think about this for a second. It stands to reason that a large flood flow with strong currents and velocity can do most channel modification. But it occurs infrequently. Equally a low or normal flow cannot do much work despite occurring frequently. It follows from this simple analysis of magnitude and frequency that a modest flow that occurs relatively frequently is they key and the discharge to which a channel will be adjusted. This is the essential conclusions of a seminal paper by Wolman and Miller in 1960 although they used applied stress rather than discharge in their analysis (Fig. 7.4). This rather mythical concept of a discharge to which a channel is adjusted is commonly referred to as the bankfull discharge. Let us explore this a little more.

Discharge (Q) is a volume of water in a unit of time, so it follows that:

$$Q = V.A$$

Where V and A are velocity and A is channel cross-sectional area. Take a hosepipe and keep the water flow constant by turning the tap on and leaving it alone. Water emerges from the other end of the hosepipe as a jet, we can make this a stronger jet (i.e., faster) by squeezing the end of the pipe and thereby reducing its cross-sectional area. We have just demonstrated the above equation. So, if discharge or more precisely bankfull discharge (Q_b) was to fall over time then the water velocity must also fall, this will cause deposition which will reducie the cross-sectional area. The link between velocity and deposition comes from the Hjulström Curve (Fig. 8.5). Equally, if we now increase Q_b then the velocity in a smaller channel must also increase causing erosion which will in time increase the cross-sectional area. In theory therefore our alluvial channel should be in equilibrium with the bankfull discharge. We can see this manifest in channel adjustments downstream of new dams. The dams reduce the discharge, and the channels often have sediment benches along their banks, a channel within a channel if you like.

- Task 8.4: What can you find out about the impact of dams on downstream reaches? Think in terms of the geomorphology and the aquatic or riparian ecology. Try to find some specific examples.

The planform geometry of alluvial rivers is also adjusted to discharge although in a more complex fashion. We can recognise three basic types of channel pattern within alluvial rivers: straight channels, meandering channels, and braided channels. These are shown as a matrix in Fig. 8.10 against various variables including discharge and stream power.

Natural straight channels are usually stable with well vegetated banks which makes them resistant to erosion. Roots are wonderful things in binding sediment and helping to resist erosion. At the other endmember we have braided channels which by contrast are usually rather unstable. They consist of multiple channels separated by bodies of sediment (bars). They tend to occur where the stream power is high, and the banks are easily eroded. Deposition within the channel is common. If we consider this for a moment, we can see how it follows from the discharge equation. As the channel widens, due to the erodible banks, the cross-section

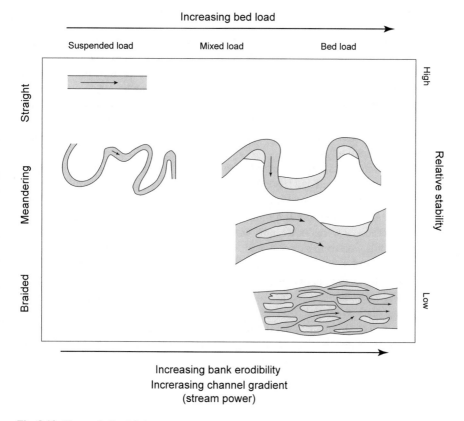

Fig. 8.10 Types of alluvial river pattern

increases and if the discharge remains the same then the velocity must fall and deposition within the channel must occur creating bars. Braided channels also occur where the proportion of bedload is high, this is highly sensitive to fluctuations in discharge and therefore seasonal falls my lead to within channel sedimentation again leading to bar formation. For these reasons braided rivers are common in arctic and arid areas where vegetation sparse and the sediment load high.

Between these endmembers we have meandering rivers which can vary in their stability, but classically migrate laterally across the floodplain (Fig. 8.11). The outside, slightly downstream, part of the meander erodes and is often associated with a steep bank and pool. The inside of the bend is associated with deposition in the form of a berm of sediment which is called a point bar and is composed of individual scroll bars. Lateral migration is a function of secondary currents, that is water flowing at right angles to the primary down stream flow.

As water moves around a bend it is flung outward under the influence of centrifugal force such that the water surface is super-elevated on the outside of the bend with respect to the inside (Fig. 8.12). In fact, this is measurable with fairly basic surveying equipment. This differential drives a flow of water across the channel towards the inside of the bed. It also drives a small current toward the outside bank which is responsible for bank erosion and ultimately the lateral movement of the meander. The pattern of current is reversed at the next bend and is more symmetrical in the reaches between bends which are often shallow and described as riffles. The secondary currents are the key to understand the evolution and movement of meanders over time (Fig. 8.12). In some rivers the sinuosity appears to be in equilibrium with the sediment and water discharge, in others it appears to be unstable and meandering is a function of self-regulation or ordering. There is huge amount of research into the morphology and dynamics of meandering but some of the essential basics of why water meanders and what controls meander morphology or stability remain elusive to universal explanation.

Channel transformation or metamorphosis has been widely studied and human intervention has provided one laboratory. In the early nineteenth and early twentieth centuries heavy metal mining hit a peak in Britain, in particular the search for lead and zinc ore led to many small mining operations in Northumbria and Mid-Wales. Typically, an adit, a horizontal shaft, would be extended from the valley floor into an adjacent hill side to intersect a vein. Ore and waste rock would be brought out along the adit and waste rock dumped if not in close by the stream or river in the valley floor. It was the geomorphological equivalent of a huge increase in available bedload and often caused channel metamorphosis with braided channels replacing meandering ones. A complex pattern of sediment storage and re-working resulted added to which the toxicity of the metal waste reduced vegetation further enhancing the erodibility of the banks. As mining ceased, most of these operations where short-lived, the river system slowly adjusted and over time the river returned to meandering pattern.

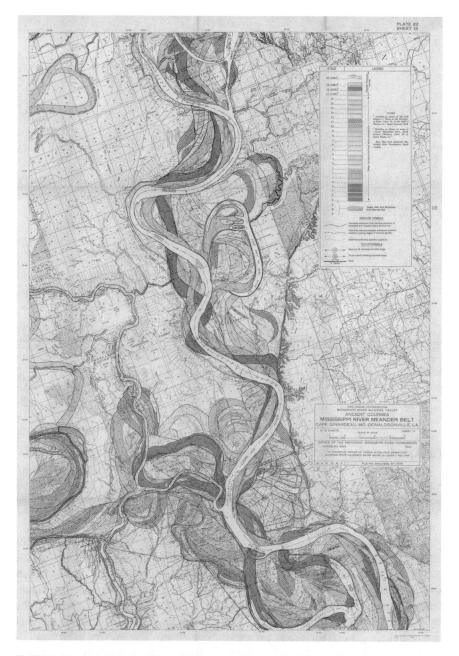

Fig. 8.11 This is one of several beautiful maps produced by the US Army Corps of Engineers for the Mississippi river which show the multiple channels and migratory patterns of the river's meanders over time. (Public domain: Geological Investigation of the Alluvial Valley of the Lower Mississippi River, published by the Army Corps of Engineers in 1944)

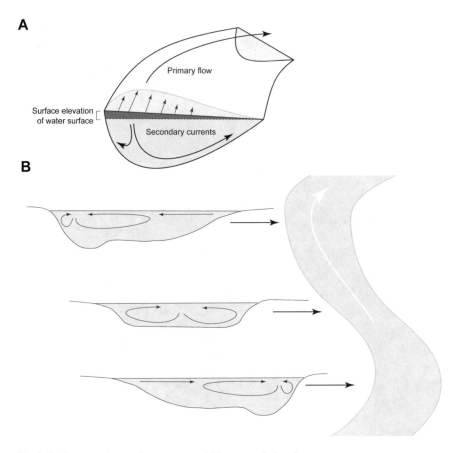

Fig. 8.12 Patterns of secondary currents within a meandering river

- Task 8.4: Take a look at Fig. 8.13. This is a hypothetical example based on research both in the Northeast of England and Mid-Wales. Write a short series of notes to explain the changes in river pattern that you see.

Climate change should also cause channel metamorphosis. As climate deteriorates vegetation may decline, causing banks to become more erodible and stream power may increase with more precipitation. In such circumstances a shift towards a more braided river system is a likely outcome. Floodplain fragments are sometimes preserved as river terraces and gravel pits or sections within them reveal the style of sedimentation, braided or meandering for example. Reconstruction of many rivers shows a general picture of channel pattern change that mirrors the major climate shifts at the end of the last glacial cycle (Fig. 8.14). Braided rivers dominate during the late Pleistocene and during the Younger Dryas with meandering rivers becoming more common in the Holocene.

On a small-scale connection between slope processes, rivers and climate have been illustrated by a series of papers published Adrian Harvey formerly of Liverpool

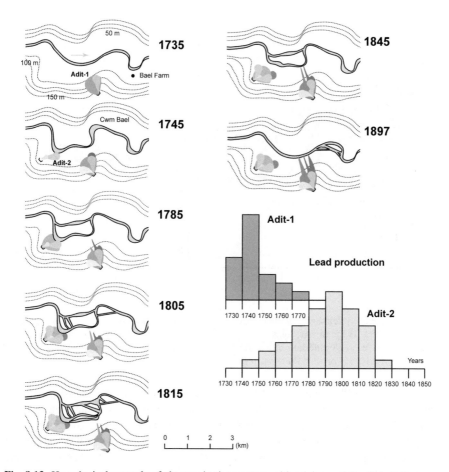

Fig. 8.13 Hypothetical example of changes in river pattern with mining activity. This links specifically to Task 8.4

University before he retired. Much of his research was undertaken in the Howgill Hills which are located just to the east and slightly south of the Lake District. He monitored rivers in the Carlingill, Langdale, and Bowderdale valleys for over 30 years. A hundred-year storm in 1982 caused widespread slope failure, gullying and re-activation of debris fans which supplied large amounts of debris to the rivers and caused channel metamorphosis. Since then, the channels have recovered and returned to more stable meandering patterns. Different valley couplings, hillslope to channel, means that the fluvial response to the storm event was different. A wide flood plain for example decouples the river from the slope. In some valleys where the coupling hill slope to channel was close the additional bed load following the storm moved downstream as a 'sediment slug' causing progressive downstream channel change long after the original storm. The fluvial response varies from valley to valley and temporally depending on the reach of river in question.

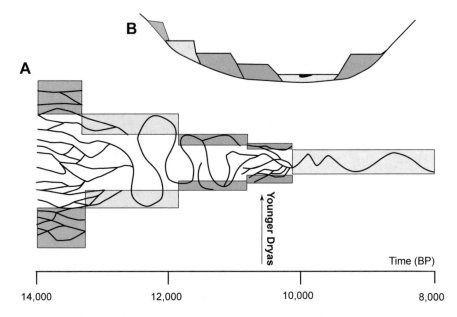

Fig. 8.14 River systems and climate. This is a theoretical model derived from several large European rivers. The plan form view of the river system through time shows how there is a transition from braided to meandering, with a brief return to a braided system during the Younger Dryas. This type of record is preserved in river terraces as shown in part (B)

We introduced something called the hundred-year storm in the last paragraph and it worth just pausing to explain what a return period, recurrence interval, or repeat interval of an event is. Recurrence interval equals (n+1)/m where n number of years on record and m is the number of recorded occurrences of the event being considered such as a flood. The theoretical return period between occurrences is the inverse of the average frequency of occurrence. For example, a 10-year flood has a 1/10 or 10% chance of being exceeded in any one year. This does not mean that a 10-year flood will happen regularly every 10 years, or only once in 10 years. It is a probability statement, which is also used to indicate size of an event. A 100-year event is by definition larger than a 10-year one.

- Task 8.5: Why not take a look at a couple of the papers by Adrian Harvey? There is a good overview in this one published in 2001 in Catena. Also, this one in 2007 focuses on how two different valleys responded to a hundred-year flood in a different way. Finally have you heard of lichenometry? This is the use of lichens to date rock surfaces Harvey et al. (1984) do exactly that using gravestones to calibrate the lichen sizes.

This lack of coupling between storm events and fluvial response makes it difficult to build up regional or countrywide relationships between climate events and river response. For example, the Little Ice Age in Medieval times should be manifest in European rivers, but in practice the record is difficult to decipher. Change in river channel morphology, activation and stabilisation of hillside fans may all bury organic matter which can be dated via radiocarbon dating. But making sense of these dates on anything but a local scale is difficult. But over the years lots of dates from lots of rivers have been accumulated as part of local studies. This data from lots of rivers across large regions has allowed the principles of 'big data' to be applied.

Rather than looking at a specific river this large integrated dataset can be used to examine the problem regionally. The key to this is to understand something about radiocarbon dates and probability. Most of these river response studies involve dating fragments of peat or soil by radiocarbon methods as mentioned above. For example, a floodplain pool, full of organic matter (plant remains, dead vertebrate and invertebrates) may be abandoned by channel movement and can be dated. Alternative a sediment fan may form at the base of a slope and extend to bury soil or peat. There are lots of different scenarios and they all record some form of channel changing event.

Now radiocarbon dates the death of the plant or animal, because at that point the carbon within it, a ratio of two isotopes C12 and C14, ceases to be replenished. At this point the radioactive isotope C14 begins to decay with a regular half-life of 5730 ± 40 years; that is the proportion of C14 decays over time in an orderly manner. If you measure the C14 that remains and know the half-life then you have an age. These age estimates are not as precise as one would always like and come with a probability; the age lies within a range. That is why radiocarbon dates are always reported with error margins. Now instead of just looking at the mean dates if one looks and more to the point combines the probability curves of multiple dates one can begin to see a bigger picture. This approach has transformed our understanding of fluvial response to climate in the last few years. It is a bit like looking at multiple radiocarbon dates and noting the overlap in error ranges and realising that collectively they say something went on at that time even though the specific means may vary from one basin to another. It smooths out the local variations and picks up the big signal. If you use this approach on data from Britain and apply various filters some key dates pop-out, such as the influence of the Little Ice Age, but significantly an event at 8200 years which corresponds to the major meltwater flood and potential shut down of the thermohaline conveyor noted in Chap. 5. Big data like this helps us gain new insight into old problems and is only possible because of the diligent work over many years by numerous scientists reconstructing and dating events in individual basins. To find out more take a look at Macklin et al. (2012).

Rock Channels

Rock channels consist of various types from those that have exposed bedrock both in the channel walls and floor, to those that have bedrock in the channel floor only to examples where it only occurs in the channel walls. In truth there is a huge amount of variety, but the essential point here is that bedrock in some way limits the development of the channel shape. To change channel form, we must invoke direct erosion, not just the remobilisation of sediment as is the case with alluvial channels.

We have shown at various points in this primer so far that rock resistance to erosion is a function not just of the absolute strength of a rock but also its total strength defining characteristics which are often defined by the weakest link in the chain. This is where rock mass strength comes into play. Used by engineers it is a total assessment of rock strength from the intact strength of a block of rock through to the spacing, density and continuity of partings (joints, bedding surfaces, faults, or any other linear weakness). The gaps or partings often provide the weakest link. We should not be surprised therefore that channel orientation and morphology is heavily influenced by such partings. There are lots of potential illustrations of this point, but a couple will serve. Figure 8.15 shows the alignment of tributaries in Navajo Sandstone in North America. They are aligned to the main joint set orientated N 20° W. The second illustrations (Fig. 8.16) show a section of the River Amazon and the alignment of the river with various faults.

Advancing beyond these simple observations is quite difficult, yet quite important since erosion of mountains chains is heavily controlled by fluvial erosion of rock channels. We can see the processes of abrasion and plucking in rock channels, but such processes are relatively slow making monitoring difficult and the environment is hard to instrument. Saltating rocks tend not to respect delicate science kit! We can transport sediment in a channel in one of four ways. We can dissolve it and transport as a solute. We can suspend the sediment in the water column. Think back to Stokes Law and settling velocity is a function of particle diameter; larger sediment grains settle fasters than small ones. In fact, a particle of clay can take over 15 hours to settle in still water. As long as the turbulence, up draft of water, is greater than the settling velocity we will have suspension. Fine-grained sediment is moved in this way. At the other extremes is sediment and rocks that are too heavy to be lifted by turbulence or fluid drag these may be rolled or slid along the bed. We refer to this as bedload and this is sensitive to velocity with slight variations causing in-channel sedimentation. Between these two extremes of suspension and bed load we have particles that can be entrained by the flow but not kept aloft within it. It is these particles that give us saltation which is of particular importance to rock erosion. Fluid drag and lift pick up these particles, but turbulence is not enough to overcome settling velocity and as a result the particle will arc back towards the bed. As it impacts on the bed it will transfer momentum to static particles resting there helping to dislodge them. It is a bit like a chain reaction; get one saltating or leaping grain and down streams it will set others in motion. The key to erosion is the impact whether due to saltation or the rolling of debris as bedload.

A new approach to understating river abrasion has developed in the last twenty years based initially on physical models and more recently on numerical simulations. This work generally goes under the term of the saltation-abrasion model. The physical modelling started with discs of rock bolted to a shaft and placed in a cylinder filled with water and sediment (abrasion mill). The disc was rotated, and the sediment/water moved relative to it with most grains moving either via bedload or salting. Two pieces of information emerged from these experiments. The first was

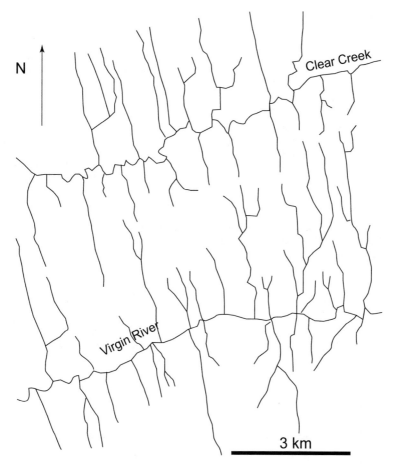

Fig. 8.15 Alignment of tributary canyons to the Virgin River and Clear Creek following pronounced joint pattern in the Navajo Sandstone east of Zion Canyon in Utah. (Gregory, H.E. 1950. Geology and geography of the Zion Park region, Utah and Arizona. US Geological Survey Professional Paper 220, Figure 96)

that little or no erosion was observed when fine-grained material was used in the experiments that carried in suspension. Erosion rates increased with increasing bedload. The second point was that as sediment load increased erosion rates initially increased as the water became better charged with sediment, but thereafter decreased because the sediment began to shield the rock from abrasion (Fig. 8.17a). The final point is that erosion rates fall with increasing rock strength (Fig. 8.17b).

One of the implications of this work is that fluvial incision into rock becomes limited quickly by the accumulation of debris and secondly the size of debris supplied by slope failure to a channel is critical. To fine and no erosion will occur to oarse and the channel will become clogged with debris.

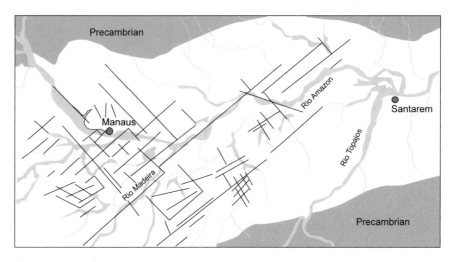

Fig. 8.16 Alignment of the River Amazon with joint sets. Faults are shown by the solid black lines. (Potter, P.E., 1978. Significance and origin of big rivers. *Journal of Geology* 8613–33, Figure 8)

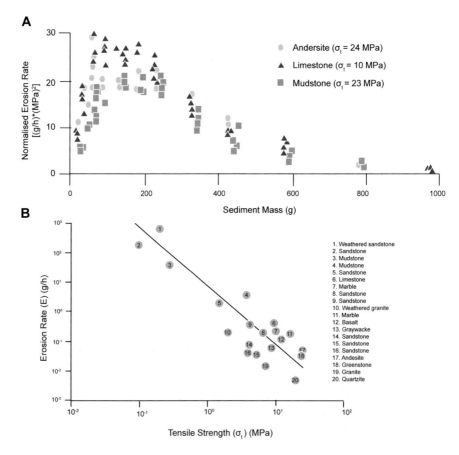

Fig. 8.17 Results from abrasion mill experiments. (Sklar, L.S., Dietrich, W.E., 2001. Sediment and rock strength controls on river incision into bedrock. *Geology* 29, 1087–1090)

Summary

There are a lot of different dimensions in this chapter from the meteorology of rain-fall through to the geomorphological evolution of landscapes under the influence of rivers. All elements are part of the hydrological cycle yet are studied by different disciplines meteorology or geomorphology and it is good to remember that it is all part of an integrated whole.

Further Reading[1]

Charlton, R. (2007). *Fundamentals of fluvial geomorphology*. Routledge.
Hamblyn, R. (2001). *The invention of clouds*. Picador.
Harvey, A. M. (2001). Coupling between hillslopes and channels in upland fluvial systems: Implications for landscape sensitivity, illustrated from the Howgill Fells, northwest England. *Catena, 42*, 225–250.
Harvey, A. M. (2007). Differential recovery from the effects of a 100-year storm: Significance of long-term hillslope–channel coupling; Howgill Fells, northwest England. *Geomorphology, 84*, 192–208.
Harvey, A. (2012). *Introducing geomorphology: A guide to landforms and processes*. Dunedin Academic Press Ltd.
Harvey, A. M., Alexander, R. W., & James, P. A. (1984). Lichens, soil development and the age of Holocene valley floor landforms: Howgill Fells, Cumbria. *Geografiska Annaler: Series A, 66*, 353–366.
Macklin, M. G., Lewin, J., & Woodward, J. C. (2012). The fluvial record of climate change. *Philosophical Transactions of the Royal Society, A370*, 2143–2172.
Selby, M. J. (1993). *Hillslope materials and processes*. Oxford University Press.
Smithson, P., Addison, K., & Atkinson, K. (2013). *Fundamentals of the physical environment*. Routledge.
Toy, T. J. (1977). Hillslope form and climate. *Geological Society of America Bulletin, 88*, 16–22.
Whiting, P. J., Bonniwell, E. C., & Matisoff, G. (2001). Depth and areal extent of sheet and rill erosion based on radionuclides in soils and suspended sediment. *Geology, 29*, 1131–1134.

---

[1] Most textbooks carry a section on rivers and my preferred one Smithson et al. (2013) *Fundamentals of the Physical Environment* does, you will also find separate chapters on rainfall and hydrology. The bits are all there not integrated as here. If you are interest in slope process there is a more detailed book written by M.J. Selby (1993) *Hillslope Materials and Processes*, published by Oxford University Press. Adrian Harvey's (2012) Introduction to Geomorphology has a few relevant sections and is accessible. There are a number of specialist textbooks on fluvial geomorphology which you could look at. Charlton (2007) *Fundamentals of Fluvial Geomorphology* Routledge is one recent one.

Chapter 9
Flowing Ice

Glaciers and ice sheets are a key component of the Earth's dynamic system both in terms of shaping the landscape and also as key elements of the climate system as we saw in Chap. 5.

In terms of their role in shaping the landscape the extent of mid-latitude ice sheets at the height of the last ice age means that a significant part of the Earth's continental area was carved and shaped by ice (32% at Last Glacial Maximum). Over the twentieth century there was a prolonged debate about the efficiency of ice as an agent of erosion with those arguing for its role as a key agent in shaping the landscape and those seeing its role as a protector. As our understanding of glacial geomorphology has evolved and both positions now appear correct as we will see later.

Despite the fact that Louis Agassiz first proposed the idea of an ice age in 1840 at a meeting of the Geological Society of London. Glacial geology is a relatively young discipline transformed by the increasing accessibility of glacial regions. Antarctica was opened up just over a hundred years ago with the grand Edwardian expeditions of Shackleton, Scott, Amundsen, and Mawson. Well into the 1960s travel to glacial regions was a time-consuming business involving ships and dog teams. Despite the accessibility of the European Alps little actually glacier monitoring, and investigation was undertaken following the pioneering work of the Professor Forbes on the Mer de Glace (Fig. 9.1). It was access to Svalbard and Iceland that was transform the discipline from the late 1960s onwards. Access to modern analogues provided a means of understanding the glacial deposits of the mid-latitude continental areas such as Britain. Before the advent of the modern analogues glacial sediments were interpreted in a more formal stratigraphic context in which tills (boulder clay) were seen simply as evidence of a glacial advance with little process-based understanding of how the sediment was actually laid down.

The challenge is well illustrated by a small coastal site on the Llyn Peninsula in North Wales called Glanllynau. This is a classic site with a three-part sequence, till

Fig. 9.1 Beautiful engraving of a glacier table on the Mer de Glace near Chamonix. From "Travels through the Alps of Savoy" (1843) by Scotsman James Forbes. The rock protects the ice below from melting thereby becoming elevated. (Public domain from "Travels through the Alps of Savoy" 1843 by James Forbes)

at the base is separated by sands and gravels from more till at the top. The traditional stratigraphic interpretation is that it represents two glacial advances separated by outwash deposits. A geologist G.S. Boulton looked at this sequence early in the 1970s fresh from an expedition to Spitsbergen the largest of the islands in the Svalbard archipelago. He used his experiences to provide a process-based interpretation in which the whole sequence was laid down by one advance coupled with ice marginal sedimentation between ridges of buried ice just like the sequences he had described from Spitsbergen (Fig. 9.2). It helped transform the interpretation of British and European glacial sequences and moved the discipline towards a process based future driven by observation at modern glaciers which throughout the latter part of the Twentieth Century became progressively more accessible. The development of remote sensing and in particular the availability of high-resolution digital elevation models has provided a second major, if less dramatic, transformation injecting more life into the discipline. So, let us back up and start with the basics how does ice flow and glaciers grow and decay?

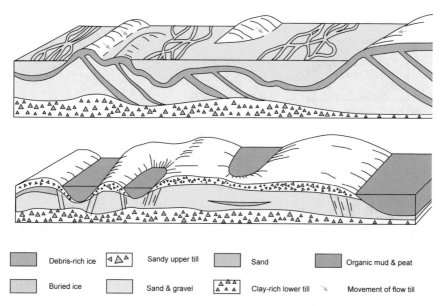

	Debris-rich ice	◁△ᴬ	Sandy upper till		Sand		Organic mud & peat
	Buried ice		Sand & gravel	△△△	Clay-rich lower till	⤺	Movement of flow till

Fig. 9.2 The sequence at Glanllynau below and the interpretation above. Debris rich ice shields the ice from melting creating a relief in which sand and gravel is deposited. Debris flows from the upstanding areas over the sand and gravel to give an upper till layer. (Boulton, G.S., 1977. A multiple till sequence formed by a Late Devensian Welsh ice-cap: Glanllynau, Gwynedd. *Cambria* 4, 10–31)

Physics of Glaciers

A quick tour of Google Earth and you can find glaciers and ice sheets of different sizes and shapes. We have the great ice sheets of Antarctica and Greenland that submerge the topography beneath them, we have ice fields like those in Norway, Svalbard and Iceland in which topography exerts some control over the ice and we have the valley glaciers of the European Alps or Alaska which sit within the valleys. Despite this range of shapes and sizes they are all governed by the same basic physics.

- Task 9.1: Find some examples of glaciers and ice sheets and note down their location, size, area, plus any other relevant information you can find out. How do they vary?

The most fundamental concept is that of mass balance, the ice budget if you prefer. For any glacier, the total volume of ice is a function of the ice that is derived from snowfall and balanced against the mass lost by melting which we refer to by the collective term ablation (Fig. 9.3a). No different than the balance in your bank account, the solvency of which is determined by income and expenditure. Any form of frozen precipitation can add to the mass of a glacier, and it is transformed slowly over time into glacier ice. Glacier ice has little air content, while snow has lots.

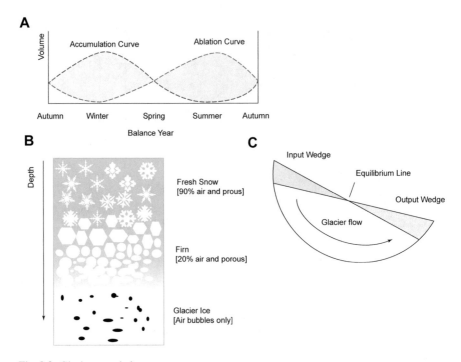

Fig. 9.3 Glacier mass balance

Burial and compaction help this transformation as does melting and re-freezing. Just think of all the arms of those pretty snowflakes that have to be crushed to remove the air between them. The intermediate state is called firn (Fig. 9.3b). The more maritime the climate (i.e., lots of snowfall and melting) the quicker this process takes place, maybe in a matter of a few years. However, in dry, cold continental locations this transformation may take hundreds of years since there is little snowfall to compact the older snow layers and little melting. Either way we add mass to our glacier over time. We lose mass by melting producing meltwater and in hyper-arid locations by sublimations which is the direct evaporation of ice. We can also loose mass by calving of icebergs if our glacier terminates in a lake or at sea.

If over time we add more mass than we lose then the overall volume of our glacier must increase, and the converse must also be true. We can shift to a positive balance (gain>loss) by increasing winter snowfall or reducing summer losses due to cooler temperatures. Today most glaciers and ice sheets are retreating due to global warming so many glaciers have a negative mass balance (gain<loss). There are some non-linear functions here which can decouple glaciers from climate.

Perhaps the best example is to think of is a fjord with a glacier lobe in it, next to a second lobe that terminates on land (Fig. 9.4, Glacier Lobes A and B). The glacier lobe in the fjord loses mass by calving icebergs. If the mass balance becomes positive over time due to an increase in snowfall and cooler summer temperatures then our ice cap, and its two ice lobes, will expand (Fig. 9.4). The land-based lobe

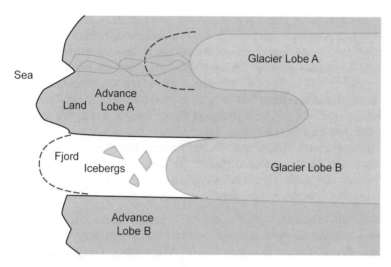

Fig. 9.4 Two ice lobes one terminating in a fjord another on land. See text for explanation of the differential advances given the same climate deterioration

advances forward proportionate to the deterioration in climate. However, the lobe in the fjord advances much more, out of proportion with the climate deterioration. This is because the number of icebergs that can calve from this lobe is determined by the width of the fjord, the calving front. The only location at which the length of the calving front can increase is beyond the confines of the fjord (Fig. 9.4) so the advance it out of proportion with climate. There are other non-linear relationships like this that may accelerate or deaccelerate glacier response to climate. This is one of the reasons that we should not assume that global warming will simply cause all glaciers and ice sheets around the world to disappear in time.

Now if we focus on the terrestrial lobe for a moment and watch it change as mass balance becomes negative over a number of years, we will see the ice front retreat. But even though the ice front is retreating, and the ice lobe is getting smaller, the ice still flows downhill towards the ice margin. Conceptually glaciers always flow downhill even when the margin is retreating upslope!

So how do glaciers flow? Certain solids behave like rheids, that is brittle when hit or stressed sharply, but ductile when stressed slowly in a sustained manner. Ice is an example of a rheid. Gravity provides the stress and the steeper the ice surface the greater the stress and therefore the faster the ice flow. Take a simplified glacier such as that in Fig. 9.3c and let us cut it in half longitudinally. The upper reaches, those higher up the mountain, will experience colder temperatures and more snow fall. In contrast those at the bottom will experience slightly warmer temperature and in theory less snowfall. The upper reaches will receive more mass via accumulation, while the lower reaches will lose more mass via ablation. Somewhere in the middle is a point at which accumulation is equal to ablation, we call this the equilibrium line, and it separates the accumulation and ablation zones. Over time the surface gradient of our glacier must increase as we add mass to the accumulation zone and remove mass from the ablation zone. As the gradient increases, shear stress rises and above a threshold ice will begin to flow. Let us hold off explaining how that flow takes place for a second and first explore one logical implication of this. In maritime areas like Alaska for example the snowfall is high and the summers mild so the surface gradient will be high. We are adding lots of mass to the accumulation zone each year and also removing lots of mass from the ablation zone. The downslope mass balance gradient will be high and in theory the glaciers in this region should flow faster. And they do at between 100 and 300 m per year.

In contrast in more continental climates where snowfall is more modest and the summers colder the mass balance gradient (i.e., the contrast between the upper and lower reaches of a glacier) is much lower. As a result, the shear stress is much lower and the glaciers flow much more slowly, perhaps as little as 5–50 m per year.

- Task 9.2: find out some typical velocities for glaciers in the Himalayas, European or New Zealand Alps and in Arctic Canada. How do they vary? Compare them to annual climate data for locations close by. What patterns do you see?

So far, we have dodged the question of how ice flows, but the simple answer is by creeping, sliding and via a deformable bed. Ice may flow by all three mechanisms or just via creep depending on how cold the ice at the bed of the glacier is. Now ice is said to be warm based if the ice at the bed is at the pressure melting point. Pressure melting point refers to the fact that as pressure increases due to the weight of the ice melting point of ice falls below 0 °C. If the temperature of the ice is below pressure melting point, then it is said to be cold-based and frozen to the glacier bed.

Creep involves the slow deformation of ice under pressure and for ice is determined by Glen's Flow Law in which strain (i.e., rate at which ice deforms) is proportional to the third power of stress. In the formulation below A is a temperature constant.

$$\dot{\epsilon} = A\tau^3$$

As we have seen shear stress (τ) is a function of the glacier gradient and ice thickness and is given by:

$$\tau = \rho g h \sin\alpha$$

In which ρ and g are the density of ice and the gravitational constant respectively, h is the ice thickness and α is the ice surface gradient. So as shear stress increases either due to thicker or steeper ice the strain (creep) increases to the power of three. Creep embraces all types of deformation from that which takes place at a crystalline level to that which occurs due to the folding and faulting of larger blocks of ice.

Now where ice is wet-based (i.e., warm based) it can also slide at the bed. A film of water at the bed helps make this possible and the greater the water pressure at the bed the easier it is for the ice to slide. We can express this as follows:

$$N = \rho g h - w$$

Where N is the load experienced which is a function of the weight of the ice (ρgh) minus the basal water pressure (w). More water pressure the less the load and the faster the sliding.

Now for a long time that is where things stood in terms of glacier movement, until our friend G.S. Boulton made another important observation. Working at Breiðamerkurjökull an outlet glacier of the Vatnajökull ice cap in Iceland he noticed that as much as 90% of the forward motion of the glacier was down to the deformation of mud beneath the glaciers bed. He conducted a number of experiments beneath the glacier to prove this using strain markers inserted into the sediment below ice tunnels. He coined the term subglacial deformation to describe this process of flow. Where ice sits on saturated, soft sediment such as clay, or sediment from an earlier glacial advance, the sediment literally moves the ice above it. To deform ice, you need between 0.5 and 1.0 bars of pressure, to deform mud you need much less, may be 0.2 bars. So, the bed begins to deform before the ice does and it is highly efficient which is one reason why glaciers moving over soft-deformable beds are usually much faster than those moving over hard-rock beds (Fig. 9.5).

One-point worth making is that if ice is cold-based and frozen to its bed it cannot slide or move by subglacial deformation so the only mechanism available is internal deformation (creep). This all helps add to the rich tapestry of different glacier velocities in different geographical regions.

Before we move on from the mechanisms of glacier flow it is worth stressing that ice does not flow at a uniform speed throughout the whole length of a glacier but reaches a maximum beneath the equilibrium line where the maximum flux of ice occurs. To help understand this take a look at Fig. 9.6. We have a simple cross-section through an ice cap (Fig. 9.6b), and we have conceptualised accumulation as a single layer and ablation as another block either side of the equilibrium line (Fig. 9.6a). So, let us start moving ice from the upper most reaches at the ice divide, in fact we will imagine a mystical dwarf with a wheelbarrow doing the work. The poor chap fills the barrow at the ice divide several times and moves these barrow loads down the glacier a distance. Once he has cleared the ice-divide of fresh accumulation he moves down stream to the new location. Here the poor chap finds that he has double the work. He has to move the first lot of snow and then the stuff that was here to start with. And so, the work goes on accumulating towards the equilibrium line at which point he has to move the whole of the accumulated layer. To be honest this is the point at which he gives up and finds a part in the Lord of the Rings. The point here is that the maximum flux of ice to be moved is below the equilibrium line and given a fixed cross-section the ice velocity must peak. Remember that discharge Q is equal to velocity times cross-section, and this holds for ice as it does for water.

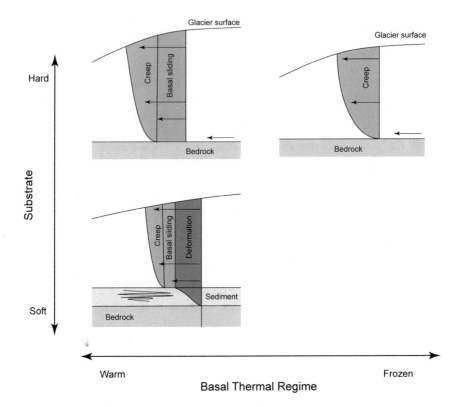

Fig. 9.5 Mechanisms of glacier flow

Glacial Erosion

Ice sheets and glaciers erode whenever their basal ice is either warm-based so they can slide and or where it is actively freezing to the bed in zones of transition between warm and cold ice. Think about it for a moment if the ice is frozen to the glacier bed it cannot slide or move, and erosion will be unlikely. In fact, at such locations the preservation of landscapes is possible. This is one reason people have argued in the past for glaciers being landscape preservers. So, the variation in basal thermal regime over both space and time will control the distribution of erosion beneath an ice sheet. We will also get preservation under the ice divide of large ice sheets since as we saw in Fig. 9.6 ice velocity is minimal at such locations irrespective of the basal thermal regime.

Where ice is wet-based and sliding we can get erosion which occurs by one of two processes either by abrasion or by plucking. Abrasion occurs where ice moves over the bed charged with rock debris either with a similar or greater hardness to the rocks of the bed. If the rock debris in the ice was softer than that of the bed it would simply be crushed. There are different models of glacial abrasion but the simplest to

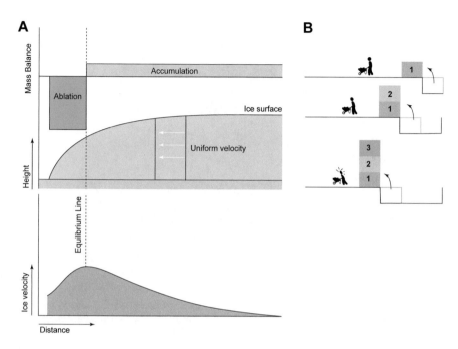

Fig. 9.6 Explanation of why glacier velocity peaks below the equilibrium line. It is zero beneath the ice divide

conceptualise involves three variable the rate of abrasion, glacial weight or load, and velocity. Observations at modern glaciers suggest that erosion rates vary from as little as 9.6 mm a^{-1} to over 36 mm a^{-1}. This data is from direct observations beneath Breiðamerkurjökull and the Glacier d' Argentière which have velocities of 9.6 and 250 m a^{-1} respectively.

If we think for a moment of a piece of sandpaper and an old door caked in paint. In removing the paint with the sandpaper, we can accelerate the process by rubbing harder and faster, but if we press to hard the grains of sand will adhere to the paint and shear from the paper. Using this analogy, the faster the glacier slides with a stone (clast) embedded in its base the faster the rate of erosion. Also, if we increase the load (i.e., press harder) on this clast we will increase abrasion up to a maximum, thereafter the increased friction caused by the load on the clast will slow it down until it finally stops and lodges on the rock surface. We have a simple abrasion model in which the key variables are glacier thickness (Load (N) = ρgh-w) minus water pressure and glacier velocity (Fig. 9.7). The product of abrasion are glacial striations (linear scratches) and a range of crescentic fractures and cracks caused by the contact of an ice set clast with the bed.

- Task 9.3: Find some facts, examples and pictures of micro-scale features associated with glacial erosion, such as striations, crescentic fractures and micro-crag and tails.

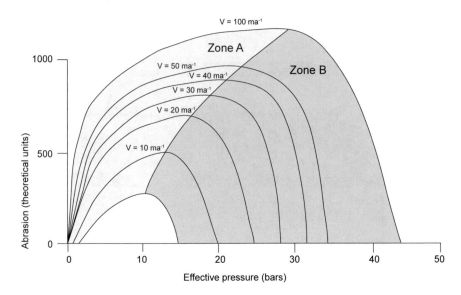

Fig. 9.7 Glacial abrasion model after Boutlon (1974). Effective pressure is the same as load (N). (Boulton, G.S. 1974. Processes and patterns of glacial erosion. In Coates, D.R., ed. *Glacial geomorphology*. New York, State University of New York, 41–87).

In addition to abrasion, we also have the process of glacial plucking, which essentially is block removal. Block removal depends on rock mass characteristics, specifically the spacing and continuity of the joints or partings. If they are closely spaced and link together then it is easy to remove a block if they are not linked together, then it is not. Plucking is favoured by the development of lee-side cavities at the base of a glacier. These develop when the ice is thin, fast flowing, and basal water pressures are high. These conditions favour a reduced basal contact pressure or load (N). Blocks are rotated out of the lee-wall and entrained by the ice during periodic cavity closures or simply by the drag of the ice. Joints may be opened and enhanced by fluctuating water pressures and by repetitive strain, but fundamentally they are largely pre-existing weaknesses in the rock mass. Figure 9.8 shows a simple conceptual model which plots rock hardness, measured via a Schmidt Hammer, and joint spacing. Where the joints are widely spaced and the rock hard, abrasion will dominate. In contrast where the rocks are densely-jointed plucking will dominate. A Schmidt Hammer by the way is a device to measure the elastic properties or strength of rock and was invented by Ernst Schmidt, a Swiss engineer. The hammer measures the rebound of a spring-loaded mass impacting against the surface of a sample. The harder the surface the greater the rebound.

There is one particular scenario that can favour intense glacial plucking and that is where there is a transition from warm, wet ice at the glacier bed to cold ice. Here two things combine, first the deceleration of the ice caused by the warm to cold transition causes compression and the faster ice rides upwards as a series of stacked layers. Think of major road smash in which cars have rear-ended slower ones in

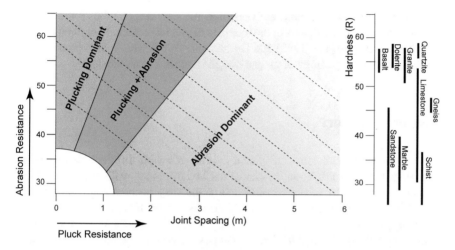

Fig. 9.8 Relationship between rock mass properties and different types of glacial erosion. (Krabbendam, M., Glasser, N.F., 2011. Glacial erosion and bedrock properties in NW Scotland: abrasion and plucking, hardness and joint spacing. *Geomorphology* 130, 74–383, Figure 9)

front to form an imbricate stack. Secondly as the wet basal ice freezes it locks on to the rock surface enhancing the entrainment of debris. So, zones of thermal transition often found at ice margins are sometimes zones of maximum plucking.

In terms of the products of glacial erosion we can view these at three scales. At a micro-scale we have striations, friction cracks and the like already discussed. At a meso-scale we have things such as roche moutonnée and whalebacks. Roche moutonnée have a plucked downstream face and a smooth, abraded up ice surface, although in practice this depends on the geometry of the joints and partings in the rock mass. At a larger scale (macro-scale) we have giant roche moutonnée and glacial troughs which are often and incorrectly referred to as U-shaped valleys. As one geography teacher once said to me, depends on your handwriting so best to avoid the term!

Glacial troughs, fjords if they are water filled, form over multiple glacial cycles and their geometry is strongly influenced by the characteristics of the rock mass. Rock that is weak and highly jointed typically give broad flat-bottomed troughs, while stronger volcanic rocks that are less well jointed are associated with narrow and deeper troughs. The network of glacial troughs is usually much more connected than a valley system that results from fluvial erosion alone. This is because ice can move between valleys via high cols which become eroded over time to connect valleys. In a lovely piece of work from the 1970s Valeria Haynes used network descriptors developed initially for the analysis of transport systems to describe the connectivity of valley systems. She demonstrated and was able to map across parts of Scotland changes in valley connectivity that appeared to reflect the intensity of glacial erosion.

- Task 9.4: Take a look at Haynes (1977) and also Sahlin et al. (2009) who both use connectivity indices to look at the intensity of glacial erosion. You might also like to look at Augustinus (1992) and Brook et al. (2004) who both explore the link between the morphology of glacial troughs and rock mass strength.

Glacier Transport

We can recognise two types of glacial debris both with different characteristics (Fig. 9.9). Debris that falls on the glacier surface from rock walls above the glacier surface is referred to as supraglacial debris and may become englacial debris if buried by ice and thereby incorporated into the glacier. It is coarse, angular, unimodal in grain size, and blocky just like any rock fall debris. This is not surprising since it derived from rock fall! The second type of debris is subglacial debris either derived by erosion or by englacial debris that reaches the bed. In contrast it is heavily modified with rounded corners and faceted sides that are often striated. Typically, it has a bimodal size distribution consisting of coarse rock (lithic) fragments and clay-silt sized clay grains produced by crushing of the lithic fragments. At a glacier terminus basal debris may be elevated by compression and melting such that it become visible on the glacier surface. The contrast between the two debris types is sufficient for glacial geologists to tell the difference and work out the history of a rock or sediment when encountered in a section. Having looked at erosion and transport the next logical step is for us to consider the products of glacial sedimentation.

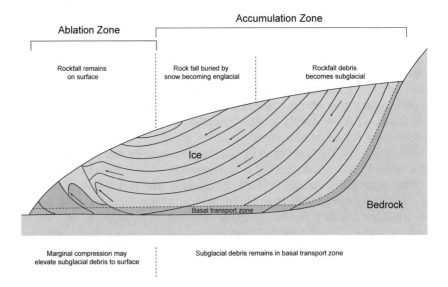

Fig. 9.9 Debris transport pathways

Depositional Landsystems

Glaciers produced lots of different types of sediments and associated landforms and these vary between those deposited on land and those deposited in water. To make sense of this and to pick out a few highlights for this chapter we will use a concept known as landsystems. A landsystem is a recurrent assemblage of landforms and sediments which are produced by a specific set of processes. We recognise just three glacial landsystems here but there are many more, they are: subglacial landsystems, supraglacial landsystems and valley glacier landsystems.

Subglacial landforms are produced, as the word implies, at the glacier sole (Fig. 9.10). The surface expression of a subglacial landsystem is streamlined landforms known as drumlins, or megaflutes. The name drumlin is derived from the Gaelic word druim ("rounded hill," or "mound") and first appeared in scientific use around 1833. The difference between a fute and a drumlin is simply the degree of elongation, with flutes being many times longer than they are wide. Drumlins vary widely in size, with lengths from 1 to 2 km with heights from 15 to 30 m, and widths from 400 to 600 m. A drumlin is one of the most studied and perhaps iconic of glacial landforms and is described as being half a boiled egg in some work or half a torpedo in others. In fact, there is a lovely analogy which equates drumline shape to eggs and relative egg size to bird that laid it (Chorley, 1959). The more elongated ones are those in which the egg size is large compared to the size of the bird; greater

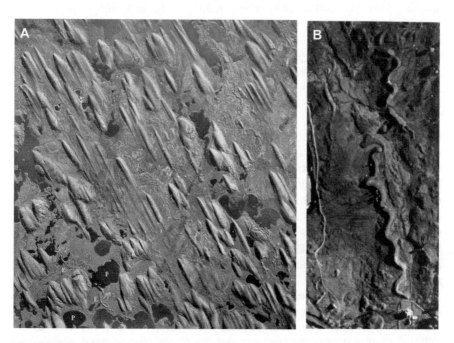

Fig. 9.10 (**a**) Aerial photograph of a drumlin field in Canada. (**b**) Aerial photograph of an esker in the Scottish Highlands

the streamlining that is needed to get it out! Like most streamlined hydrofoil-like shapes they tend to be blunt up-flow, taping down-flow.

Drumlins may be made from (1) glacial till, (2) rock, or (3) sand and gravel. This is the point we need to introduce you to a till. By definition till is a sediment that is brought together by the direct agency of a glacier. Technically speaking where a glacier deposits its sediment in water the resulting sediment is no longer a till, yet it may look exactly the same. This is why most people tend to use the non-genetic term diamicton (diamict) these days. If you are interested this comes from the Greek dia- meaning through and meiktós meaning mixed. This is simply a term for sediment with a large mix of grain-sizes, which is what a till usually has (although not always!). Traditionally glacial geologists used to define a range of till types using names like lodgement, melt-out and deformation till. This was based on the idea that sediment lodged below a moving glacier, smeared if you like onto the bed, would have properties different from one where the basal debris melted out from the sole of the glacier or where the glacier was moving over soft-sediment and deforming it. However, with the advent of micromorphology this view changed, and most researchers now simply talk about subglacial till. So, what is micromorphology? Basically, it is a thin-section through the sediment viewed under a microscope. A steel tin is hammered into an *in-situ* block of sediment, extracted, and then impregnated with resin back in the lab. This solidified block is then sectioned in the same way you section a rock for analysis. The point here is that careful analysis of lots of thin-section from different types of basal till all show that they have been deformed to some extent. The distinction between the till types began to disappear.

So, today the practice is to consider all basal till to be the same, formed by lodging of sediment, melting of basal debris and the assimilation of soft debris over which the glacier is flowing. This sediment is then deformed and results in a diverse range of basal tills which have all been tectonically modified and over-consolidated by the weight of ice. So, our subglacial landsystem is dominated by subglacial till overlying an eroded rock surface. Now back to drumlins!

So, drumlins can have a range of internal constituents from rock cores smeared by subglacial till to cores of sand and gravel again smeared by till, and finally some drumlins are completely composed of subglacial till. The current prevailing view is that drumlins are the product of subglacial deformation. You can thank G.S. Boulton for this idea, he was a truly inspirational glacial geologist. His basic idea was that sediment moves in a deforming layer. This sediment is then smeared or deformed around different types of obstacles, a knob of rock, a fold of stiffer till, or a body of sand and gravel. The point about the sand and gravel is that it drains more freely than the deforming layer. If the porewater of the sediment is lower there is less to facilitate its deformation and it is stiffer. The greater the porewater pressure the more the water pushes the individual grains apart allowing them to move past one another. So, think about a meltwater portal at an ice margin. Sand and gravel build up around this and downstream of the portal. If the ice now advances over this ground the area either side of the meltwater portal, made of less well-drained till, will deform, but the area of sand and gravel will not. This resilient core will be eroded and shaped by the deforming layer either side to create a drumlin. Till properties may also vary spatially in terms of grain-size and porewater content. Stiffer areas may be deformed by folding creating obstacles whereas softer areas will deform and move rapidly around them. In this way subglacial deformation around different types of obstacles create drumlins with different compositions. There are some mathematical studies which have shown that random instabilities within a deforming layer may generate drumlins all by themselves and this is an area of active research. One of the consequences of drumlins being formed in this way is that they are orientated in the direction of deformation which is normally the same as ice flow. They can be used therefore to map ice flow directions.

The main constituent of our subglacial landsystem is likely to be subglacial till, the surface of which may be drumlinised, or at least streamlined (Fig. 9.11). There may also be packages, or lens, of sand and gravel deposited by subglacial channels. Where these are found at the surface, they may form eskers which are sinuous ridges of sand and gravel that mark the location of subglacial meltwater streams (Fig. 9.10b). The word esker derives from the Irish word Irish word escir, which means "ridge or elevation, especially one separating two plains or depressed surfaces".

This streamlined landsystem contrasts with that generated in a supraglacial setting. Supraglacial simply means on the surface. In certain situations, large amounts of debris may build-up on the glacier surface (Fig. 9.12). This is common where a glacier terminus encountered a reverse bedrock slope which causes deacceleration and rising debris layers. Think about a traffic accident where the cars in front slow rapidly but those behind keep speeding forward. The resulting compression causes thickening of the layer of cars. This is also true were basal ice transitions from warm and wet to cold. Basal sediment layers are frozen onto the glacier sole at this

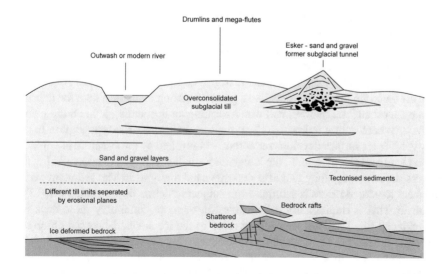

Fig. 9.11 Elements of a subglacial landsystem

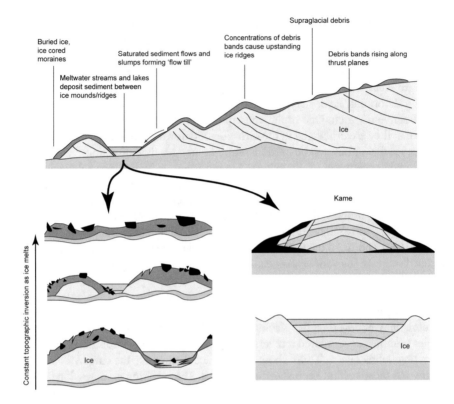

Fig. 9.12 Supraglacial landsystem

junction (warm to frozen) and large volumes of sediment may be generated, the
glacier will also slow at this point causing the debris bands to rise. The net result is
an ice marginal zone that can be thick with debris.

Initially a thin debris layer may accelerate melting, because dark surface absorbs
more heat, but as that layer thickens it begins to insulate the surface. A sediment
with all those air-filled voids acts like a giant insulating blanket. Where the debris is
thickest least surface melting will occur forming an upstanding area on the glacier
surface. Where debris is finest surface melting will lower the topography. In this
way the glacier surface develops relief that is controlled by the structure of englacial
debris in the glacier margin. In the low point we will get rivers and lakes all of which
lowers the ice topography and also deposits sand and gravel. The sediment on the
adjacent ice-cored ridges is saturated and subject to debris flow and slumping. This
result in what is traditionally called flow till, or more commonly these days just
supraglacial till. The properties reflect those of the ice slope, debris saturation and
therefore the fluidity of the resulting flow. Pebbles are aligned not to glacier flow but
to the local ice surface slope. This whole environment is extremely unstable.
Slumping from ice cored ridges may reveal fresh ice for melting causing the topography to invert. High points may become low point and the constant churn of

sediment creates a real mix of sediment properties. Where there is greater stability in the supraglacial landscape a kame may be generated (Fig. 9.12). This is a ridge or mound made of sand and gravel, the lower point of the former river becomes a high point. The term kame is a variant of comb (kame, or kaim is the Old Scottish word for comb), which has the meaning "crest" and was used in 1874 as a geological term by Thomas Jamieson.

The final type of landsystem is that associated with valley glaciers. Figure 9.13 shows some of the key components of which we can recognise four main landform sets. The first of these is the lateral moraine of which you can see two broad types in Fig. 9.13. On the left-hand side of the image there is the eroded trace of a lateral moraine. The main debris source for this moraine when formed was probably the scree or talus slope you can see on the valley side. This combined with supraglacial debris being dumped off the side of the glacier builds a ridge which when the buttressing support of the ice is removed rapidly degrades to a mere trace. On the right-hand side the lateral moraine here is primarily built as a dump moraine with material sliding from the ice margin. Frontal moraines come in many different types and can result from sustained advance or from seasonal ice-marginal fluctuations. Common at many temperate glacier margins are small push moraines formed each winter by a minor seasonal advance. These ridges are generated in a number of different ways from the simple sweep of the ice front gathering debris as it advances to more complex mechanism involving the freezing of sediment slabs to the thin ice of the margin in winter. These slabs are ripped-up and stacked as the ice advances. Seasonal

Fig. 9.13 Glacial valley landsystem. Numbers are: 1 = frontal moraines, 2 = lateral moraines, 3 = medial moraines, and 4 = glacio-fluvial outwash

push moraines like this form because ablation slows in winter, but the ice still flows. Think about this for a moment. The ice marginal position is a function of forward flow of ice and rate of melting. If melting exceeds flow, as it does in the summer, then the ice margin will move backwards up valley. In winter rates of ablation (melting) are low, but the ice still flows forward and therefore the margin must advance.

For much of the Twentieth Century mass balance of glaciers in Norway and Iceland have been negative. On average over a whole year more ice is lost than is gained by snowfall and as result glacier margins have been retreating. This retreat is marked by a succession of annual push moraines formed each winter. The up valley spacing between each of these winter moraines is a function of summer air temperature and the warmer the summer the bigger the distance. There are some nice studies which have explored this link.

Larger frontal moraines may be formed by more sustained advances associated with positive mass balance, but the evidence is usually lost as the ice advances over these moraines. One particular non-climatic cause of a rapid advance is the glacier surge. About 10% of all glaciers exhibit surge-type behaviour and these glaciers are mainly located in Alaska, Svalbard, and Iceland. These glaciers show quiescent behaviour most of the time, but at regular intervals surge forward at rates of up to 800 m in a year. They push everything in front of them creating large tectonic moraines (Fig. 9.14). They are active for a couple of years and then the marginal advance stagnates, and the glacier goes back into quiescence. The cause of this binge-purge behaviour is not well understood but involves the switching of basal friction between super-slip and super-stick. They are much studied by glacial geologists because of the large and interesting landforms they generate. The small moraines in front of our valley glacier each mark the position of the margin, but a surge-moraine is a bit like a rucked-up rug in front of sliding child with multiple ridges in advance of the ice margin.

This brief canter through some of the depositional landforms and landsystems produced by glaciers and ice sheets gives a small glimpse of the range of landforms that are generated. One of the advantages of take a landsystem approach is that one

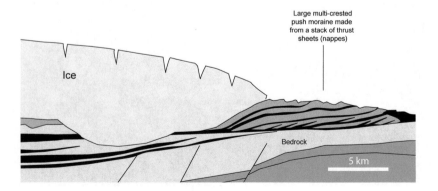

Fig. 9.14 Large tectonic moraine produced by a sustained glacier advance or surge

can map out the occurrence of different landform assemblages to reconstruct ice sheets that have long since melted. In fact, much of the landscape of Britain, North America and Fennoscandinavia is simply a palimpsest of glacial landsystems. By mapping these landsystems you can reconstruct the geometry and behaviour at the height of the Last Glacial Maximum. The reconstruction of past glaciers in this way is something which is known as palaeoglaciology.

Further Reading[1]

Augustinus, P. C. (1992). The influence of rock mass strength on glacial valley cross-profile mor-phometry: A case study from the Southern Alps, New Zealand. *Earth Surface Processes and Landforms, 17*, 39–51.

Bennett, M. R., & Glasser, N. F. (2009). *Glacial geology* (2nd ed.). Wiley.

Brook, M. S., Kirkbride, M. P., & Brock, B. W. (2004). Rock strength and development of gla-cial valley morphology in the Scottish Highlands and northwest Iceland. *Geografiska Annaler: Series A, 86*, 225–234.

Chorley, R. J. (1959). The shape of drumlins. *Journal of Glaciology, 3*, 339–344.

Haynes, V. M. (1977). The modification of valley patterns by ice-sheet activity. *Geografiska Annaler: Series A, 59*, 195–207.

Sahlin, E. A., et al. (2009). Connectivity analyses of valley patterns indicate preservation of a preglacial fluvial valley system in the Dyfi basin, Wales. *Proceedings of the Geologists' Association, 120*, 245–255.

Smithson, P., Addison, K., & Atkinson, K. (2013). *Fundamentals of the physical environment* (4th ed.). Routledge.

---

[1] Smithson et al. (2013) *Fundamentals of the Physical Environment* has a couple of good chapters on glaciers and glacial geomorphology. The best source for a more detailed introduction is Bennett and Glasser (2009) *Glacial Geology* published by John Wiley and Sons.

Chapter 10
Coastal Processes

Planet Earth has a lot of coastlines. Estimates vary depending on the scale at which the measurements are made the figure lies somewhere between 1.16 and 1.63 million kilometres. The variation is sometimes referred to as the coastal paradox and is due to fractal properties. If you measure the UK's coastline at an interval of 200 km the total length is 2400 km but repeat the measurement at 50 km and you gain an additional 1000 km of length. The true figure is closer to 12,000 km. According to the World Factbook, prepared by the CIA no less, Canada has the most at over 202, 080 km and the UK has 12, 429 km. Figure 10.1 shows a cartogram originally published in the *Geographical Magazine* which distorts a world map by coastal length. Look how little coastline Africa has and how much larger the UK looks.

Coastlines are biologically important, and they have played a major role in the evolution of life. In particular tides are the key, causing periodic wetting and drying of coastal areas facilitating biodiversity and in the past the movement of life from the oceans to the land. Coasts are also a big part of managing the interaction of humans with their environment given that 44% of the World's population lives within 50 km of the coast and some place the figure even higher.

The macro-scale geography of the coastline is determined by plate tectonics and the contrast between a collision coast and a trailing edge or passive margin marked. Think of the western seaboard of the Americas, particularly South America, where the Pacific Plate descends beneath the continental plate of South America. The coastline is compressed, narrow, with high mountains behind and an offshore trench. Deep water lies close to the shore and the rivers although powerful are short and relatively light in sediment load. Contrast with much of the coastline of Africa. Here there is no active plate boundary (except for the Red Sea), the continental shelf is wide, coastal waters shallow and there are large rivers delivering lots of sediment to the coast. The eastern seaboard of South America is another example of a passive

M. R. Bennett, *Our Dynamic Earth: A Primer*,
https://doi.org/10.1007/978-3-030-90351-0_10

Fig. 10.1 Cartogram showing a world map distorted by coastal length. (Public domain: http://geographical.co.uk/places/mapping/item/1912-coast-lines)

margin. The macro-scale distribution of erosion and depositional landforms along the coast reflects this plate tectonic control at a high order.

• Task 10.1: Look up a paper by Inman and Norstrom (1971) which sets out the of plate tectonic in the macro-scale character of coasts. Look at the distribution of some coastal landform types such as deltas. Glaeser (1978) also looks at the distribution of barrier beaches in light of plate tectonics and Dickinson and Valloni (1980) use it to show how marine sediment vary in type around the world.

Importance of Tides

A tide is a distortion in the shape of one body induced by the gravitational pull of another nearby object. In the case of the Earth the gravitational pull of the Moon and the Sun distort the ocean surface. The gravitational attraction of both the Sun and the Moon work either in opposition, or together but let us ignore the Sun for now.

The gravitational pull of the Moon leads to two bulges in the ocean surface one directly below the Moon and another on the opposite side. The Earth rotates beneath these two bulges giving two high tides in every 24 h (Fig. 10.2). The fact that the Earth has an axial tilt means that these two bulges are not highest at the equator. Take a droid standing on the Earth's surface, lost and alone, and let us image that sea level was so high as to flood everything. Our drowning droids would rotate with the Earth and pass under two bulges in water level of unequal height (Fig. 10.2). Their precise location would determine the relative size of these two high water levels. That is why tidal regimes around the World tend to fall in to one of three categories diurnal, semidiurnal, or mixed. Also note that timing of each high tide progresses by

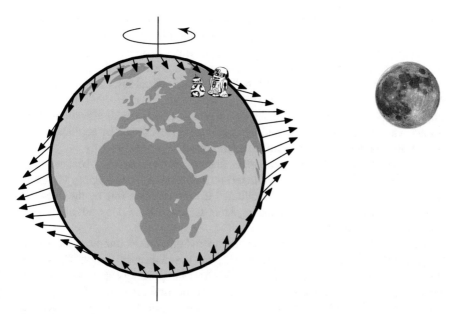

Fig. 10.2 Tidal forces cause two high tides each day. Because of the axial tilt of the Earth, shown here correct to the vertical, the Star Wars droids on the Earth's surface experience two different tidal levels, one higher than the other as they rotated beneath the two bulges. If they move elsewhere on the plant the relative size of the two tides will vary

50 min in 24 h because the Moon is not stationary and rotates around the Earth. It therefore takes a little longer to return to the same spot directly under the Moon because of this. Let us rescue the droids by lowering sea level to normal levels. Changing the water level within the oceans will complicated the tidal pattern locally due to the interaction of the tidal wave (bulges) with the local basin geometry and this adds a lot of local colours. The Severn Bore in the UK is one example of this local colour.

- Task 10.2: Look up the tidal regime of different locations around the World. Try and find tidal graphs and compare these to find locations with different tidal regimes. Note down a few contrasting examples.

The big challenge is to understand why there are two bulges, one beneath the Moon and one on the opposite side of the Earth. It is the explanation for this that most textbooks get wrong. Most textbooks talk about the gravitational pull of the Moon stretching the water surface. In fact, it is squeezed not stretched, and this is amplified by the sheer-size of the oceans which cover some 70% of the Earth's surface. It is one of the reasons why we do not get tides in lakes or in your cup of coffee. Figure 10.2 shows the vectors of tidal force, see how they are directed inwards at the poles and toward the axis of maximum gravitational attraction between the Moon and the Earth. These forces pushing the water surface upwards. But why two bulges? Well, the surface of the Earth is also experiencing gravitational forces which pull and move it slightly towards the Moon. The second bulge is caused by the residual force, that is the forces acting on the water minus those on the Earth.

So far, we have ignored the Sun and it works either in opposition to the Moon or in concert with it. During a lunar cycle, the relative position of the Moon and Sun change. Sometimes they are aligned so that the gravitational forces increase the size of the two water bulges giving us Spring Tides, while at other times they work in opposition at right angles to one another thereby reducing tidal elevations (Neap Tides). The magnitude of Neap and Spring tides is also affected by equinoxes when the distance between Sun and Earth are minimised.

Tides are some of the most important features of our coasts not because of the geomorphological work they do but in terms of creating a varied habitat. The movement of life from the oceans was facilitated if not made possible by the tides. The chances are if there is life on another planet then it will be one with at least one Moon and therefore tides.

There is one other important consideration and that is tidal duration. Tides mean that locally sea level is always changing. Now the highest of Spring Tides occur infrequently and therefore at that level a cliff for example only experience the direct impact of waves occasionally. Some water levels occur more frequently is the point here, those for example at intertidal levels or close to mean high and low water neap tidal levels. If we sum all the time spent at different levels, we have a tidal duration curve, which will vary with the tidal regime (Fig. 10.3) but will effectively determine the levels at which we get most water contact. This is important for understanding ecological zonation on cliffs as well as modelling areas of peak erosion.

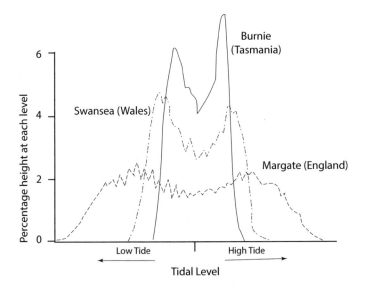

Fig. 10.3 Tidal duration curves, the higher the peak the longer the time that water is at that height for. (Carr, A.P., Graff, J., 1982. The tidal immersion factor and shore platform development: discussion. *Transactions of the Institute of British Geographers*, 240–245. Figure 2)

Making Waves

Waves are formed by the movement of energy through repeated localised oscillations of matter (in our case water) about some mean or resting position. In water the
oscillation of water transfers energy in the form of wave-form that normally progresses, but the water does not progress. So as wind blows over a water surface
momentum is transferred wind to water and the resulting energy forms a progress-
ing wave. Waves can progress or be stationary. Stand on a hilltop and yell 'I am
geographer get me out of here' and the wave will progress into the distance. Now
find a cave and repeat this and the sound will echo around you as a stationary or
trapped wave. Another example of a trapped wave is the sting on a guitar when
plucked it move to the bridge and back constantly until the energy is all consumed.
Think of the wave as lots of molecules of water orbiting on the spot passing with
energy from one orbit passing to an adjacent one and so on. Why labour this complex definition? Well, it is important to emphasise that waves only become geomorphologically significant when they break, and this energy is converted to currents of
flowing water.

The wavelength of waves created in deep water varies with the wind and the
resulting waves are a chaotic collection of resulting waves moving at different
speeds, with different wavelengths and frequencies. Where these waves sync-up,
crest-on-crest and trough-on-trough, they combine, where they are out of sync, they
subtract from one another. Provided that the base of the oscillating water column
that makes up the wave does not touch the seabed then waves are pretty efficiently
energy transfer systems and can progress over long distances. As they do so the different frequencies may begin to sort themselves out and become more consistent
(swell waves) although they are always prone to being 'mixed-up' again by local
storms and winds. This sorting is due to the fact that wave velocity (C, celerity) is a
ratio of wavelength (L, distance between two crests or troughs) and the period (T,
number of wave crests passing a point in a unit of time). This gives us the classic
wave equation:

$$C = L / T$$

The stronger the wind the more energy is transferred to a wave and the more energy
it will have. This energy manifests itself in wave height (H) and the relationship
between the two is given by:

$$E = \frac{1}{8} pgH^2$$

The formation of wind waves includes an initial slow growth phase associated with
the formation of new waves on a calm water surface. This is followed by a rapid
growth phase where the increasing roughness of the sea surface makes the energy
transfer from wind to waves more efficient. As a wave field develops there is an

increase in the wave height and wave period. Wave growth is not continuous, however, and water turbulence in the wave (white-capping) acts counter to this growth and dissipates energy. As sea waves radiate away from the location where they were generated, the waves with the greatest wavelength (or period) travel fastest and those with the shortest travel slowest. This causes them to become dispersed, according to their wavelength, into organised swell. We generally refer to the normal wave regime as the wave climate. That is the waves that occur most commonly at a site over the long-term and not just those found on the day you visited.

All is set until the waves begin to feel the bottom which occur at about half their wavelength (L/2) at which point friction begins to cause the wave to modify and shoal. As waves shoal, or 'begin to feel' the seabed or beach foot, their properties change. First of all, the friction slows the velocity of the wave form. Now just like in a traffic jam if the wave ahead slows the gap to the wave behind, still in slightly deeper water and moving a little faster, must fall. In essence the wavelength shortens. Due to the conservation of energy if the wave form slows kinetic energy falls and is transformed into potential energy by raising the wave height. So, as the wave shoals its wavelength falls, its height increases, and it follows that the steepness of the wave L/H must increase. Now while the celerity of the wave form falls the orbital velocity of the water molecules remains the same and this will ultimately lead to the wave breaking.

There is one other important aspect here. Think of a line of students, arms linked and moving along on skateboards, one aside the other. If they all travel at the same speed the line will stay straight, but if one end of the line begins to slow then the line will tend to swing or pivot around that point. The line has undergone refraction. The same is true of a wave shoaling, sections running over a shallower bed will slow faster than those over deeper areas and the wave crest when seen in plan will change shape with some bits advancing more than other bits. Let us represent our wave crest in planform, that is seen from above by a drone, as series of points, dots on a dotted line if you like. We have a height at each dot and therefore a wave energy. As the crest of the wave changes in length due to refraction the gap between some of our dots shortens and in other places grows. So effectively does the wave energy per unit length of a wave crest. This concentrates wave energy in some locations, while spreading it out in others. This is an important coastal property and also one that will change as sea levels fall or raise both with tides and over longer time periods.

Ultimately as a wave moves inshore it will break when the wave has slowed so much that the velocity of the orbiting water molecule (u) exceeds the forward velocity (C) and the wave form collapses. This tends to occur when the ratio of water depth to wave height lies somewhere between 0.6 and 1.2 with a mean value of around 0.78. This means that low waves run father into shallow water than tall ones. The beach angle is also important with waves breaking closer inshore where the beach is steep and further out where they are flat. There is something called the breaker coefficient (B) which defines this as follows:

$$B = \frac{H}{LS^2}$$

Here S stands for beach slope, L for wavelength and the H for wave height. We can use this to define different type of breaking waves which have different characteristics and break either close inshore or further offshore (Fig. 10.4). This is important because the more inshore a wave breaks the more asymmetrical the onshore and offshore component of water flow is and will be once it has broken.

Current asymmetry is an interesting and important concept, but not the easiest to grasp. Think about two water pistols, with the same capacity and barrel diameter. One of the pistols shoots water onshore, the other fires it offshore. Press the trigger for the same length of time, with the same force and the same amount of water is jetted onshore as offshore in the same amount of time. The two currents are symmetrical. Now if we increase the pressure forcing the water through the barrel in the case of the onshore barrel, we get a higher velocity, but the water pistol will run out more quickly. If we now put less pressure on offshore pistol trigger, then it will jet out at a lower velocity over a longer time period. The two currents are now

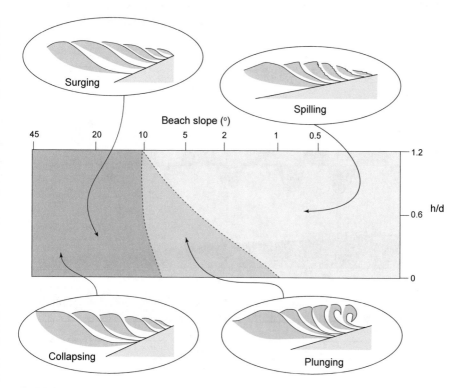

Fig. 10.4 Relationship between beach slope, water depth and wave height which defines for basic types of breaking wave. (Smithson, P., Addison, K., Atkinson, K., 2008. Fundamentals of the physical environment. Third edition Routledge, p358)

asymmetrical. In each case the same volume of water is moved just with different velocities and over different periods of time. This is the essence of shore normal current asymmetry. Water moves onshore quickly and over a short amount of time, but it moves offshore slowly and therefore over a longer period of time to move the same water volume. This asymmetry develops more strongly if the wave breaks inshore, while a wave that breaks further offshore gives more symmetrical current flows. Spilling breakers which break offshore give symmetrical shore normal currents, while surging waves give currents that are more asymmetrical (Fig. 10.5).

There is one final aspect here that is worth noting and that is waves can be reflected. Just as light waves are reflected by mirrors water waves can be reflected by cliffs and even beaches. The steeper the beach the more energy is reflected. Some of these reflected waves may become trapped in the surf zone as edge waves. Edge waves have crests which lie perpendicular to the beach and the incoming waves. Basically, they are the expression, hard to see mind, of trapped energy and the interaction of Edge waves and incoming waves can cause the wave set up (height) to vary along shore. Wave set up is the amount by which the water level is elevated or setdown on average by the incoming waves. If the waves are higher the setup is greater and the currents that flow along the beach will be faster. Edge waves can cause the height of an incoming wave to vary along its length in a cyclic fashion. This causes the set up to vary and may induce along shore currents with water flowing from

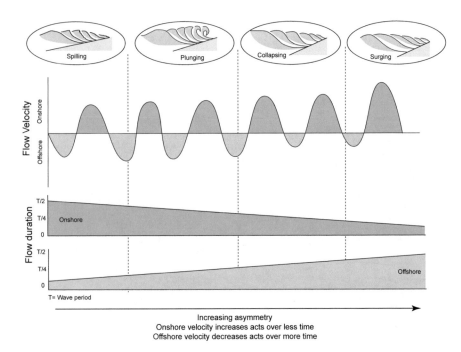

Fig. 10.5 Different levels of wave asymmetry with breaker type and position on shore. The beater symbol represents slope in the breaker coefficient equation. (Pethick (1984) Coastal Geomorphology, Arnold)

areas of higher set up to lower set up; effectively downhill. These nearshore circulation cells are important in explaining some, although not necessarily all, types of rhythmic landforms on beaches.

We now have enough information to be able to examine some geomorphological systems found on the coast. There is a huge range of different types of system along the coast from biological ones such as coral reefs, to estuaries to rock cliffs such as those found in Cornwall. We need to be selective here and focus on just two of these systems namely those associated with beaches and those associated with rock cliffs and shore platforms.

Dancing Beaches

A beach is defined as an agglomeration of cohesionless sediment. Since cohesion is a property of fine clay and silt particles (<64 um) beaches are by definition made of sand, gravel, or cobbles. The point here is that the component particles can move easily with the currents acting on the beach. A mudflat in an estuary cannot do this due cohesion. If you go to the beach regularly you will see that it changes its morphology almost on a daily basis. Sometimes it may be steep and other occasions flat and the landforms on it change. May be there are low waters bars on some occasions and on other occasions rhythmic beach cusps. The point here is that a beach is dynamic and constantly adjusting to the wave regime. When the wave climate is severe, with high storm waves the beach profile tends to be flat and under more gently regular swell conditions steeper. In exposed coastal locations, with a long fetch, receiving storm waves most days the beach will always be flat, in more sheltered locations the beach may be steeper. Fetch by the way is the distance travelled by wind across open water; the larger the fetch the more dominant will the waves from that direction be and potentially larger. Grain size also plays a part here, coarse beaches tend on the whole to be steeper than fine sand ones. Beaches are able, by natural feedback systems, to adapt to the wave energy they receive is the upshot of this and we refer to this as morphodynamics (Fig. 10.6).

The question is how and why? The why is simple to accommodate the incident energy. A beach deals with incident energy in one of two ways it can dissipate it, or it can reflect it. Dissipation involves water turbulence and waves breaking offshore as spilling breakers is the key. A flat beach with offshore bars enhances this process and the water turbulence consumes energy. Reflection is favoured by a steep beach in which the waves get close in shore to be reflected by the beach face. Ideally, we can envisage a morphological cycle in which a storm draws down the beach moving sediment offshore to create a flat beach profile dissipating the energy. In quieter swell conditions sediment moves onshore creating a steeper beach which reflects the incident energy. Now if our beach is located in an exposed location then it will tend to occupy the dissipative end of this continuum, while in a shelter location it may occupy the reflective end being drawn down occasionally by storms. The beach

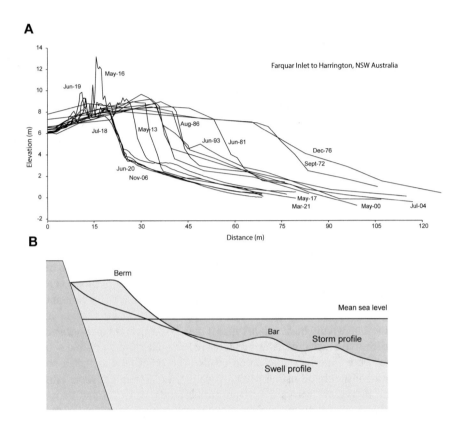

Fig. 10.6 Beach morphodynamics. Part A shows real data from New South Wales in Australia, not the beach profile variation both over the years and between the seasons. Flatter during winter months (i.e., northern hemisphere summer). Part B shows a schematic representation of the variation between storm and swell profiles. (The data used in part A is from the NSW Beach Profile database www.nswbpd.wrl.unsw.edu.au/photogrammetry/nsw/)

morphology is attempting to find an equilibrium state in relation to the wave energy which is never stable enough for it to achieve this state long term.

The question is how? It is easy to slip into phraseology that implies that the beach has a brain and 'is trying' to find this equilibrium state. Beaches do not have brains, so the process is a natural feedback of some sort. Wave asymmetry provides a conceptual way of explaining this mechanism. As we saw in Fig. 8.5 sediment entrainment by a water current is a function of grain-size and the transport threshold on a beach is determined by the velocity of the currents and the mean size of the grains on the beach. Let us take a scenario with symmetrical onshore and offshore current, the velocity and duration of both currents is the same and we have equal amounts of onshore and offshore transport. However as lower wave conditions prevail waves break further onshore, and the resulting currents are more asymmetrical. In this case the transport threshold for a given beach may only be exceeded by the faster, but shorter, on shore current. The offshore current does not exceed the transport threshold even though it has a longer duration. As a result, sediment moves on shore steepening the beach and accentuating the predominance of onshore sediment transport.

Now if a storm were to blow up large waves get close inshore to start with and will by asymmetrical. With more energy the transport threshold is exceeded for both the onshore and offshore currents, but sediment moves predominantly offshore because the offshore current has greater duration. This flattens the beach but is ultimately limited by the restoration of a more symmetrical current regime. Conceptually at least we have a mechanism therefore to explain the adjustment of a beach to wave energy. We can integrate this into rhythmic topography such as cusp and crescentic bars to gain a complete 3D morphodynamic picture.

In general terms the steeper a beach the shorter the wavelength of edge waves generated by reflection and therefore the shorter the spacing between areas of enhanced wave setup. The combination of edge waves and incident waves cause regular along shore variation in wave height. Where the wave height is slightly higher the wave setup is greater. These points on the wave break to yield a greater runup distance depositing coarser material as a cusp (Fig. 10.7). Finer sediment is then washed back in a small circulation between these points. Cusps tend to occur on coarse steep beaches under stable swell conditions.

Now flatter beaches tend to reflect less energy and what is reflected gives edge waves that typically have a longer wavelength. These alter the wave height on the incoming waves in the same way as the shorter ones but the distance between high points along a wave is greater. In this case rather than cusps, larger rhythmically spaced bars result. These can have a range of different forms, but the key feature is that they are evenly spaced along a beach. The nearshore circulation cell is also larger and often associated with strong offshore current known as rip currents (Fig. 10.8). We can integrate shore normal profile variation with long shore morphology to create a complete morphodynamic model for a specific beach (Fig. 10.9). Note that these models, while sharing generic components, tend to be beach specific.

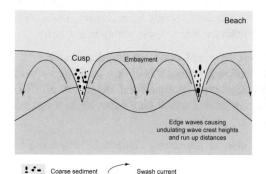

Fig. 10.7 Edge waves and cusp formation. (Photograph with permission of RHughes5)

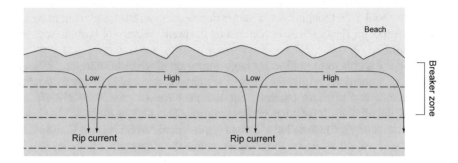

Fig. 10.8 Near shore circulation cell. The wave setup is greater between the rip currents and water flows from these highs to the areas between which are set-down thereby creating the circulation cell

The dynamic nature of beaches, and their ability to change with incident energy, makes them excellent natural sea defences. A crude analogy of my fist and a wall will reinforce this point. Without a cushion my fist might dent the wall, with a cushion to mould and shape around the fist the wall would be protected. The plumper the cushion the better the defence. The same is true with a beach the better nourished it is the better the defence. Nourishment depends on the sediment budget, how much material is being delivered to the coast by cliff erosion, fluvial transport, and via longshore transport and how much is being lost by erosion.

To help understand this it has become increasingly vogue to talk about natural coastal cells or compartment. These are normally separated by prominent coastal headlands or broad estuaries which form compartment boundaries impervious to longshore sediment transport. Within each natural cell or compartment it is possible to keep a budget and thereby effectively monitor the health of a beach. Coastal engineering can have a major impact on the budget of a cell. For example, building seawalls prevents erosion and sediment supply, equally longshore groins may limit longshore drift and dredging and/or beach nourishment may either remove or add sediment to a compartment. In short, action at one location in a sediment cell may have a downdrift impact somewhere else. Increasingly coastal management has to accommodate and work within an understanding of these complex interactions.

- Task 10.3: Chesil beach has been described as one of the most heroic feats of natural engineering. Find out some more about this 18 km long barrier beach. Bray (1997) uses a sediment budget type approach and Bennett et al. (2009) examine its internal structure using ground penetrating radar. Make a summary of at least one of these papers.

Swash alignment of beaches is also an interesting characteristic and one that leads to longshore drift and such features as spits and bars covering the mouths of estuaries. This characteristic is beautifully illustrated by a classic conceptual model from one of the best books on coastal geomorphology written by Pethick (1984;

Fig. 10.9 Example of a
morphodynamic model for
a typical beach

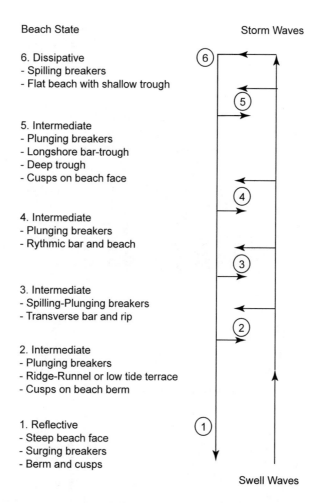

Beach State

6. Dissipative
- Spilling breakers
- Flat beach with shallow trough

5. Intermediate
- Plunging breakers
- Longshore bar-trough
- Deep trough
- Cusps on beach face

4. Intermediate
- Plunging breakers
- Rythmic bar and beach

3. Intermediate
- Spilling-Plunging breakers
- Transverse bar and rip

2. Intermediate
- Plunging breakers
- Ridge-Runnel or low tide terrace
- Cusps on beach berm

1. Reflective
- Steep beach face
- Surging breakers
- Berm and cusps

Storm Waves

Swell Waves

Fig. 10.10). We start with a beach out of phase with the incoming waves. Note the refraction around the headland which concentrates energy here and spreads it out in the bay. The angle the waves make with the shoreline is maximum between the headland and the bay and therefore it is here that longshore transport is at a peak. The result is that the headland erodes, and the bay receives the resulting sediment and over time the whole system becomes aligned to the orientation of the incoming waves and we get a so-called swash aligned beach. In practice the incoming waves may not be stable enough in terms of direction to ever reach such an equilibrium state and also sea level fluctuates changing the refraction pattern, but it is an ideal.

The distribution of wave energy has a first order control on the distribution of coastal landforms. For example, salt marshes and mudflats occur in lower energy areas, while beaches and cliffs in higher energy ones. Interestingly the location of these 'energy zones' may change with sea level rise due to changing patterns of refraction.

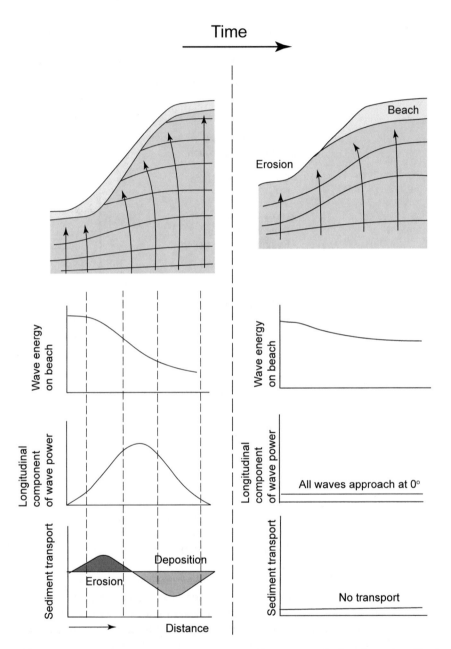

Fig. 10.10 Simple model to explain how beaches should become swash aligned over time. That is in planform the beach should follow the shape of the prevailing wave crests. (Pethick (1984) Coastal Geomorphology, Arnold)

- Task 10.4: Pethick (2001) examines the impact of rising sea level on the location of different coastal energy regimes and shows how small rises can have big impacts. This paper is also interesting since many famous and important bird conservation areas are currently located in low energy areas which will become subject to erosion with future sea level rise. If we had not fixed their boundaries so they could move with the lateral shifts in energy they would not be at risk. This is one of the consequences of conservation policies such as those in the UK that are boundary based.

Cliffs, Shore Platforms and Erosion

Cliffs and shore platforms are like an old married couple, usually seen together. The platform may not be visible, buried beneath a beach or submerged below the water line, but they are usually there. Equally the cliff may be modest in stature and the platform wide but again it is usually present. Together they are the manifestation of coastal erosion, and one only needs to tour the coastline of Devon or Cornwall to see some spectacular examples. It is however the much lower cliffs of places like Holderness and Suffolk in the UK which are often only a few metres high that are eroding fastest. We can divide cliffs generally into those composed of soft-rocks such as Eocene clays (e.g., London Clay), glacial tills and the like which are distinct from those composed of hard-rock like the folded sandstones, siltstones, and granites of SW England. The cliffs to the east of Bournemouth are all soft-rock examples, while those of Purbeck are from the other end of the spectrum with the chalk stacks of Old Harry's Rocks being an example.

Understanding why and how a piece of steep coastline looks the way it does is about understanding the relative interplay of slope processes like those discussed in Chap. 7 with marine processes. We can define two end-members, one is the classic cliff-platform combination (or a stable coastal slope) which is a supply limited system. That is the input from rockfall, or slope processes is less than the transport efficiency of the sea at that point. Waves are able to remove the debris supplied keeping the cliff steep or the slope stable (Fig. 10.11). At the other end of the spectrum is a system in which the supply of debris completely swamps the coastal transport system, and the slope is reduced. Over time, however, it may steepen again as the coastal system clears the debris steepening the slope which causes another large failure. The point here is that in the latter case it is all about the slope process and less about coastal processes. In the former coastal processes have a bigger role to play in shaping the geometry of the coastal slope, although rockfall and slope processes still play an important part.

The other key factor here is off course geology and in particular the characteristics of the rock mass. Joints, faults, and bedding surfaces are all important lines of weakness, and the geometry of a cliff is usually determined by the location of such partings. Take a look at Fig. 10.12 which shows a picture a small coastal cave west

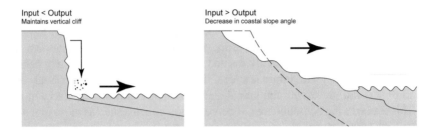

Fig. 10.11 The morphology of a coastal slope in relation to the rate of debris supply and the rate of coastal transport

Fig. 10.12 Sea cave west of Porthleven in Cornwall. Note how the geometry of the faults and joints give the cave its detailed form

of Porthleven in Cornwall. Note how the small faults on the left-hand side of the roof control the shape. The cave is also drained by a linear feature in the foreground which also follows a joint line. The jagged edges of the cave are formed by the individual bedding horizons which form steps. While this has been eroded by the sea the geometry of the cave and its presence at this location is all down to line of weakness in the rock mass.

So, what are these processes of erosion? Well, they are basically the same as those in a river, abrasion and plucking. Plucking or block removal depends initially on the rock mass properties and in particular the continuity of joints. Joint propagation is a more important process on the coast, however. Breaking waves force water and air into fractures and if done repeatedly may lead to fracture propagation and failure. A Japanese scientist Tsuguo Sunamura did a lovely set of laboratory experiments during the 1970s using a flume and cement cliff. He showed amongst other things that a wave breaking on the cliff achieved more erosion than one that broke just before. The dynamic force of the water (+ air) was more effective. Erosion is going to concentrate therefore at locations where the tidal duration is greatest; that is those point on a cliff where water level occurs more commonly (Fig. 10.3). It is a crude analogy but think of a builder with a power drill standing before a wall. They now move the drill move up and down vertically from a crouch to fully erect with the drill impacting the wall. The bit in the middle, around their waist, receives more pases with the drill than the two extremes. Abrasion can be effective as Fig. 10.13 shows, here a pothole has been eroded into a shore platform by the swirling of trapped pebbles.

As with rivers there is a relationship between the build-up of debris on a shore platform and the effectiveness of abrasion. I once shocked a lecture room full of students by asking 'had any one ever been in a riot?" I went on to say that a rioter without any bricks or stones (ammunition) at their feet could not do much harm, to many stones, or to big a barricade, and their aim would be impaired. You need just the right amount of ammunition and protection to be an effective stone thrower! The same is true of a beach resting on the shore platform. No beach and there will be no abrasive for the waves to throw, to much of a beach and the wave energy is absorbed by it. Maximum erosion is achieved with a small beach resting on the platform. Tsuguo Sunamura illustrated this beautifully with one of his cement cliff and laboratory experiments (Fig. 10.14).

The experimental cliff in Fig. 10.14 looks a bit odd since being cement the cliff tends not to collapse (no joints) so erosion is manifest by the notch. In a normal

Fig. 10.13 A coastal pothole scoured out by the rotation of the trapped pebbles

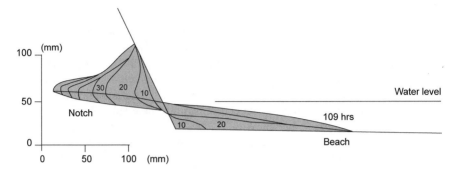

Fig. 10.14 Experimental data showing the role of a beach in first accelerating rates of cliff erosion and then in reducing it. (Modified from Sunamura, T., 1977. A relationship between wave-induced cliff erosion and erosive force of waves. The Journal of Geology, 85(5), pp.613–618. Figure 1)

setting the cliff would tend to collapse. In tropical seas the presence of a notch is much more common especially on carbonate rocks. This is created by solution but also by the work of lots of boring and grazing marine animals. On the Aldabra Atoll for example it has been estimated that a third to half of all the coastal erosion is down to grazing animals.

- Task 10.5: If you are interested coastal notches there is a nice and accessible review paper by Alan Trenhaile published in 2015 in Earth Science Reviews. Take a look and make some notes.

Bioerosion of this sort is not just restricted to tropical waters however and we can see it much closer to home. Figure 10.15 shows a section of shore platform at Kimmeridge just along the Jurassic Coast from Bournemouth. The sandy clay bedrock has lots of joints infilled by silt forming networks. A variety of marine organisms attack this finer sediment to create niches which grow to full length cracks. These cracks widen and the area between them is then hollowed out by bioerosion and the surface as a whole is gradually lowered. This is a classic example of bioengineering in which the marine organisms are creating and modifying their habitat. This can be deduced by a thought process known as ergodic reasoning or time-space substitution. Sounds a bit as if this came from an episode of Dr. Who but in fact it simply says that a process occurs at different stages in different places and can be set in a temporal framework by comparing different places.

- Task 10.6: There is a paper by Andrews and Williams (2000) that shows the impact of limpet erosion on chalk cliffs in Sussex. They estimate that the average limpet ingests 4.9 g of chalk each year which equates to an erosion rate of about 0.15 mm per year where they colonize modestly but in dense patches this may rise to 0.49 mm a year. Check it out is a cool study.

Initial colonizers of rock surfaces, including bacteria, algae, lichen, and fungi all of which may modify the surface and promote colonization by gastropods, chitons, echinoids, and other grazing organisms which scrape the rock surface as they feed.

Fig. 10.15 Bioerosion and ecoengineering at Kimmeridge, Dorset

There are at least 12 invertebrate phyla have members that can bore directly into substrate including barnacles, sipunculoid/polychaete worms, bivalve molluscs, and Clionid sponges.

Summary

We have just stuck of toe into the water really when looking at coastal process and geomorphology and there are many different aspects from the physical process discussed here to the biological ones associated with ecoengineering. A significant portion of the World's population lives on the coast and managing the interaction of human resource needs, with those of the natural environment and its conservation is a challenge.

Further Reading[1]

Andrews, C., & Williams, R. B. G. (2000). Limpet erosion of chalk shore platforms in Southeast England. *Earth Surface Processes and Landforms, 25*, 1371–1381.

Bennett, M. R., Cassidy, N. J., & Pile, J. (2009). Internal structure of a barrier beach as revealed by ground penetrating radar (GPR): Chesil beach, UK. *Geomorphology, 104*, 218–229.

Bray, M. J. (1997). Episodic shingle supply and the modified development of Chesil Beach, England. *Journal of Coastal Research*, 1035–1049.

Dickinson, W. R., & Valloni, R. (1980). Plate settings and provenance of sands in modern ocean basins. *Geology, 8*, 82–86.

Glaeser, J. D. (1978). Global distribution of Barrier Islands in terms of tectonic setting. *Journal of Geology, 86*, 283–297.

Inman, D. L., & Nordstrom, C. E. (1971). On the tectonic and morphologic classification of coasts. *Journal of Geology, 79*, 1–21.

Pethick, J. (1984). *An introduction to coastal geomorphology.* Arnold.

Pethick, J. (2001). Coastal management and sea-level rise. *Catena, 42*, 307–322.

Smithson, P., Addison, K., & Atkinson, K. (2013). *Fundamentals of the physical environment* (4th ed.). Routledge.

Trenhaile, A. (2016). Rocky coasts—their role as depositional environments. *Earth Science Reviews, 159*, 1–13.

Trenhaile, A. S. (2015). Coastal notches: Their morphology, formation, and function. *Earth Science Reviews, 150*, 285–304.

---

[1] Smithson et al. (2013) *Fundamentals of the Physical Environment* has a couple of good chapters on oceans and the coast as do must other physical geography textbooks. If you want more detailed information, I cannot recommend more highly the book by Pethick (1984) *An Introduction to Coast Geomorphology* published by Arnold. It may not be in print anymore, but there are copies in the library. Do not be put off by the fact that I used it as an undergraduate in the 1980s it is a classic and by far the best introduction to coastal geomorphology. I also direct you to a review paper by Trenhaile (2016) which covers rock coasts rather well.

Chapter 11
Landscapes Around Us

Many of you who take one of my classes will end up working in the UK in whatever capacity, some of you will work overseas. It is hard for me to cater for you all, but in this chapter, I want to look at the landscape in a more holistic fashion and apply some of the knowledge we have gained in the previous chapters to specific places. I am going focus on the UK, before moving on to have a brief look at the Amazon Rainforest.

British Landscapes

British landscapes are rich in history and has a representative selection of rocks from most geological periods ranging from the Precambrian to the Quaternary. Its ecosystem is not natural however and has been modified by humans and their ancestor since they first entered these lands over a million years ago. In fact, what we try and conserve in areas of Outstanding Natural Beauty, as Sites of Special Scientific Interest, or National Parks is not natural at all, but has been systematically created by human hands. You might well ask what are we actually conserving and why? The truth is that we are conserving some *artificial* moment in time that we have decided is 'natural' not the original landscape which is for every changing.

I once sat with my father on the summit of Moelwyn Mawr in North Wales, as I have done many times since, looking down one way to slate quarries of Blaenau Ffestiniog and the other way toward Trawsfynydd and the concrete reactor houses of Magnox reactor (Fig. 11.1). Immediately below us was Llyn Stwlan one of many pump storage schemes created to store the electricity generated by the nuclear power plant. What is natural in this landscape I asked my father? The slate quarries were carefully excluded from the boundary of the Snowdonia National Park, yet they are as Welsh as anything in the landscape and now a World Heritage Site.

© The Author(s), under exclusive license to Springer Nature Switzerland AG 2022
M. R. Bennett, *Our Dynamic Earth: A Primer*,
https://doi.org/10.1007/978-3-030-90351-0_11

Fig. 11.1 Trawsfynydd nuclear power plan. Building started in 1959 and the plant became active in 1965 running till 1991. The plan is to restore the area to its pre-1959 state by 2083. Note the dam of Llyn Stwlan high on the hillside, one of several pump storage-schemes in the area. When National Power Grid was in surplus power from Trawsfynydd was used to pump water up to these high lakes so it could be released, generating hydroelectricity, when power demand surged. https:// en.wikipedia.org/wiki/Trawsfynydd_nuclear_power_station#/media/File:Trawsfynydd_Nuclear_ Power_Plant.jpg

Having grown up in Snowdonia I have always accepted them as part of the landscape and its history. But when I looked at the concrete reactor houses, they looked stark and out of place and to modern, I questioned if they should not be removed in time. Yet my father pointed out that they are now part of the evolving landscape so why remove them? Why remove them but not the slate quarries? Where is the baseline for conservation of this landscape or any landscape for that matter?

When I was growing up Sunday night TV was dominated by the BBC's adaptation of *All Creatures Great and Small*, recently remade by Channel-5 in the UK. The original depicted an idealised version of the North Yorkshire Moors from the 1930s which was exactly how the general public expected the moors to be when they visited. In fact, this 1930s idle was in danger of becoming the conservation goal of the associated National Park for a time. You cannot fossilise a landscape in this way, nor should you try, change, extinction, adaptation, and migration are normal. What is my point here?

Well, the philosophy of conservation is complex and dependant on the time scale through which you view it. Landscapes (and climate) change slowly all the time, it is natural. We risk being blinkered when we see the world on human time-scales of just a few generations and should always remember that it is but a mere flash in the history of our planet (in fact 30 years is $6.5 \times 10^{-7}\%$ of the Earth's total age). Plants and wildlife come and go like a flickering light, but landscape on which those light plays made of rock endures and changes more slowly. This is one of the main messages in this primer: look at the Earth with true temporal perspective and not with the *arrogance* of a human who is but a brief visitor to this planet and wants to decry any change upon it. The natural world changes and you can't stop that change nor should you try, just by being alive you impact the planet in some way, but this does not mean that we should not minimise those impacts where and when we can. Good stewardship demands nothing less.

So, let us strip away the vegetation and look at the basic skeletal landscape of Britain. That landscape is for the most part a glacial one hewn and smoothed during the last 2.8 million years during the Quaternary. At some point ice covered everything north of a line between London and Bristol and even south of this line there was a small ice cap on Dartmoor. Where no ice covered the land, intense periglacial processes operated to modify the surface. Few landscape features in Britain, if any, survive from earlier periods in the Cenozoic. Periglacial by the way is the term given to processes and associated landforms that form under conditions where the ground is either seasonally or permanently frozen. Think Siberia or the Northwest Territories of Canada.

If we look at a climate record for the Quaternary such as that in Figs. 5.4 and 5.13, we see lots of glacial and interglacial cycles and must conclude that the UK has been glaciated on multiple occasions. If one looks at the deep glacial troughs of the Scottish Highlands you can see the work of these multiple glacial cycles, but in truth we have little other evidence of these cycles in the landscape. This is because a glacial advance erodes the evidence of earlier advances and makes it hard to work out what happened before.

Take Fig. 11.2 for example a simple cross section through a hypothetical continent with mountains to the left and sea to the right. A succession of three ice sheets grow and decay across this landscape one after the other, each driven by a glacial cycle. The first glacial cycle erodes the lands surface and smears it with glacial debris and deposits sediment offshore. This is followed by the next ice sheet which erodes the landscape and removes almost all of the evidence of the first cycle. The final and largest ice sheet follows during the next glacial cycle and the landscape is

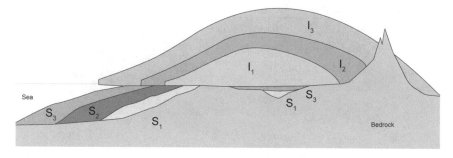

Fig. 11.2 Illustration of the imperfections of the terrestrial glacial record. S_1 to S_3 are the sediments deposited by ice sheets I_1 to I_3

again scoured. All land-based evidence of ice sheet two is removed except for the offshore sediments, because no ice sheet follows ice sheet three its glacial sediment is left on land. So, the terrestrial glacial record consists of a fragment of ice sheet one and evidence of ice sheet three, but no evidence of ice sheet two. The offshore record is more complete but less easy to access. The point here is that the land record is fragmentary and incomplete, and so it is with the UK.

When I was at school the conventional wisdom was that there was evidence of just three ice sheets in Britain one during the last three most recent glacial cycles. These were in order of old to young the Anglian Glacial, Wolstonian Glacial and the most recent was the Devensian the last phase of which is referred to as the LGM or Last Glacial Maximum (or Dimlington Stadial). The Devensian reached its maximum around 20,000 years ago and is constrained in age by a type-site at Dimilington on the Holderness coast. Here organic fragments occur below and above the glacial till and have been dated to bracket age of the glacial. The Anglian glaciation (antepenultimate) is the one that got all the way down to the London to Bristol line and occurred some 300,000 years ago. The complex one is the Wolstonian (200,000 years go), originally defined by an ice front crossing the Midlands that involved an ice-dammed lake just outside Birmingham and was attributed to the penultimate glacial. But during the 1980s this evidence was reinterpreted, and the ice front was assigned to a retreat stage of the Anglian (Fig. 11.3). Evidence of a whole ice sheet was removed at single stroke. The extent of the ice during this penultimate glacial cycle is not known, presumably because it lies inside the limit of the Devensian ice sheet and has been eroded. It simply reflects the changing nature of science, new evidence and interpretations come to light all the time and can radically change our view of the past.

The Anglian Glacial and associated ice sheet was responsible for a massive transformation of the British landscape. Prior to this time the rivers of the midlands tended to flow eastwards from the Welsh Boarders towards the North Sea. Perhaps

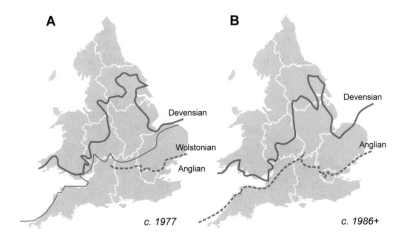

Fig. 11.3 Changing picture of glacial ice limits in the UK

the most important of these was the Proto-Soar which flowed from the Vale of Evesham toward Leicester, across the Fens (the Fens did not exist at the time) into Norfolk/Suffolk and from there into the North Sea around Lowestoft. The River Thames went nowhere near the site that would be London in the future, instead from Oxford it ran via Bury St Edmunds into Suffolk (Fig. 11.4). Then came the Anglian Glacial and a huge ice sheet pushed south down the North Sea and a large lobe of ice eroded the Fen Basin. This ice lobe ripped through the river sediments of the Proto-Soar and pushed on toward London diverting the River Thames close to its current course. A radical reorientation of lowland Britain was revealed when the ice retreated (Fig. 11.4). Instead of being dominated by west-east flowing rivers the main rivers flowed north south, the Severn south toward Bristol, the Thames south past oxford and then east along what is now the Thames Corridor. The landscape of lowland Britain, the landscape of Shakespeare, of the Midlands and of the Fens is really rather young. The evidence of this glacial event is subdued, largely to be found in gravel pits, because the periglacial conditions of two subsequent glacial periods have worked to erode and subdue them.

Inside the limits of the LGM the evidence is much fresher. In the last 10 years or so, with the advent of high-resolution digital elevation models (DEMs) a group of UK glacial geologists set out to map in detail the deposits and landforms of the Devensian and produce a definitive glacial map of the UK.

- Task 11.1 take a look at Evans et al. (2005) which reviews the evidence that went into the glacial map of the UK or Clark et al. (2004) which is the map itself.

The study of past ice sheets and their reconstruction is the subject of palaeoglaciology and it is possible using satellite data, DEMs and fieldwork to reconstruct an ancient glacier. Why would you bother? Well mainly because by looking at the behaviour of past glaciers we can predict how current ice sheets may change in

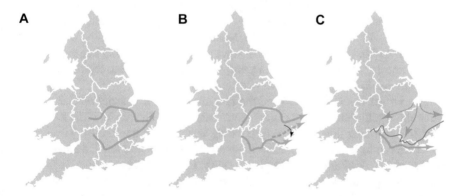

Fig. 11.4 Transformation of the lowland areas of Britain during the Anglian Glaciation. As shown in Part A the River Thames ran across Suffolk and was joined by the proto-River Soar. The wash basin did not exist. Over time the River Thames shifted southwards via the Vale of St Albans (Part B). The Anglian ice sheet eroded the Wash basin and terminated in north London as shown in Part C. The River Thames was diverted to its current course

response to future climate change. What data do we need to reconstruct an ancient ice sheet? Well, we need to know its maximum horizontal and vertical extent and we need to know its pattern of ice flow which will tell us where the ice divides once lay. Different types of ice-marginal landscape may tell us something about the extent, not only the maximum extent but the pattern of retreat as the ice sheet decayed at the end of a glacial cycles. Ice-marginal moraines, kames, outwash fans, and the extent of glacial sediment may all help. Dating of these deposits especially by cosmogenic exposure dating can also help constrain what landforms belong to which ice sheet. It can get quite complex where the one set of landforms are superimposed over another. In terms of terms of vertical extent, we can use something known as a trimline. Trimlines are lines (sometimes zones) where there is a transition on a mountain side between glacial features below and periglacial features above. Basically, a transition from terrain that was ice covered and is eroded to terrain that was never ice covered and stuck up above the ice as a nunatak.

Now if we know the direction of ice movement then we can deduce where the centres of mass (ice divides) of an ice sheet were. Ice flows from high point towards the margins. We can use erosional landforms such as striations and roch moutonée to deduced ice flow directions, but these tend to reflect quite local flow patterns. For large scale flow directions, we need drumlins and megadrumlins. We saw in Chap. 9 how these form by subglacial deformation and therefore should be aligned to the direction of ice flow. By mapping them we can deduced ice flow directions. In the 1960s and 1970s this was done for the whole of the North American or Laurentide Ice Sheet. Using thousands of aerial photographs Canadian glacial geologist mapped all the landforms and produced an amazing reconstruction. But they missed something rather important.

In the late 1980s a young postgraduate student Chris Clark was recovering from a mountaineering accident. He was a contemporary of mine and we were both working on different projects under G.S. Boulton in Edinburgh. Chris saw something that no one else had seen; that the drumlins and megaflutes cross-cut one another. Where the Canadians had mapped converging flow lines, Chris saw ones that cross-cut one another. It was an observation that has transformed our understanding of ice sheet behaviour and has allowed us to reconstruct past ice sheets with a huge amount of fidelity not just at their maximum extent, but through time. So, backing up for a moment why do drumlins cross-cut one another?

The answer lies in Fig. 11.5 and the fact that ice sheet velocity reaches a peak under the equilibrium line but is negligible under an ice divide. So, if the ice divide were to move over time, perhaps responding to change in the Rossby Waves as is now believed to be the case for the Laurentide Ice Sheet, we can have landforms preserved. Think of a series of ice divides (Fig. 11.5) moving over time. In the first position we get radial ice flow, that is outwards from the ice divide. Peak deformation will occur below the equilibrium line. Now if the ice divide were to move over time the pattern of ice flow would become reorientated. Under the ice divide where the ice velocity is negligible the drumlins from the first ice flow direction would be preserved, while below the equilibrium line they would be destroyed and reorientated. But between these two point the ice flow evidence would be superimposed. It

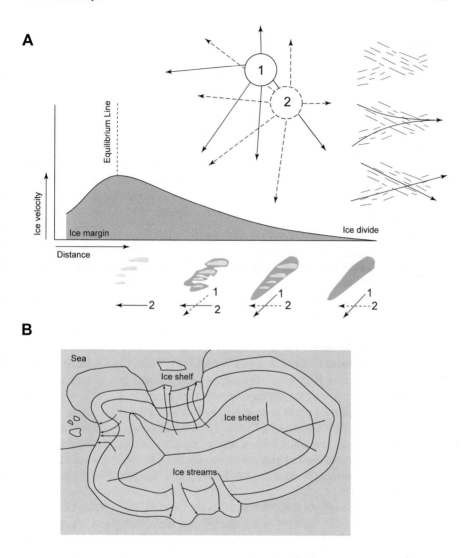

Fig. 11.5 The development of cross-cut drumlins and megaflutes. (**a**) shows how the movement of an ice centre may create changing patterns of ice flow. Divide one moves over time to a new position (2). (**b**) The faint dashed lines are the orientation of individual drumlins. These where traditionally interpreted as converging flow lines, but when re-examined they can be seen as two cross-cutting flows. (**c**) Deformation with ice velocity which reaches a peak under the equilibrium line. (**d**) Reconstructed hypothetical ice sheet. Note the ice streams which may concentrate ice flow directions. (Modified from: Clark, C.D., 1993. Mega-scale glacial lineations and cross-cutting ice-flow landforms. *Earth Surface Processes and Landforms* 18, 1–29, Figure 13)

is a process that does not just need movement of the ice divide, the turning on and off of ice streams (corridors of fast flow) during deglaciation may also cause cross-cut patterns. This simple observation, taken to its logical extent, has transformed our understanding of past ice sheets and their dynamics.

- Task 11.2, today there is widespread concern about the stability of both the Antarctic and Greenland ice sheets, if they melt sea level could rise by more than 60 m. One issue is the stability of ice stream, corridors of faster flowing ice. Palaeoglaciology can really help here. Stokes et al. (2016) reported the behaviour of ice streams during the decay of the Laurentide Ice Sheet. Rather than finding them to be unstable elements their number decreased with deglaciation. Take a look at this paper since it is a powerful justification for the study of ancient long-gone ice sheets in dealing with modern environmental issues.

The moraines and sediments of the Devensian or LGM ice sheet in the UK are not the last glacial sediments to have been deposited in Britain's mountains. Following the LGM the British Ice Sheet began to decay as climate warmed. Ice may have almost completely gone, at least from lowland areas when suddenly things got cold again during the Younger Dryas. You may recall this was a short sharp return to glacial conditions at the end of the last glacial cycle caused perhaps by the outpouring of meltwater from lakes dammed up by the Laurentide ice sheet in North America. Whatever the cause things got icy again for a couple of thousand years. Glaciers returned to the uplands of Wales, Lake District and a small ice field returned to the Scottish Highlands. The degree to which this was a readvance or a fresh advance varies across the country. In Wales for example the ice had gone, but in the Scottish Highlands it might have still lurked in some valley heads. The evidence is fresh only 10,000 years old, fresher than that associated with the LGM, and the event occurred at time when Milkanvotich orbital cycles show a warming, and these ice bodies formed the last glaciers to exist in the UK. So, there has been a lot of interest and research ever since the late 1960s. Their interpretation has been subject to a lot of debate.

The traditional paradigm was established in the early 1970s by Brian Sissons, he supervised lots of PhD students at Edinburgh and each was given a part of the Highlands to map. Many of these students went on, and still do in some cases, hold influential positions in Quaternary Science in the UK. They used a morpostratigraphic method. Inside the limits of the Younger Dryas glaciers there was lots of moraines which they described as 'chaotic' from the perspective of someone on the ground field mapping. This so called 'hummocky moraine' was used to define the limits of these glaciers, hence the term morphostratigraphic units. That is a landform assemblage of one age, the age being the Younger Dryas. They did not attempt to explain this assemblage, suggesting that it was evidence of rapid glacier stagnation. At the time evidence was emerging of just how fast climate warmed at the end of the Younger Dryas, so the idea of what was referred to as mass or areal stagnation fitted. Climate warmed rapidly and the glaciers died or stagnated in their tracks leaving a bunch of debris as hummocks. Their aim was always to determine the maximum ice limit especially for small valley or cirque (corrie/cwm) glaciers. The reason for this was there was a simple empirical way of determining what the

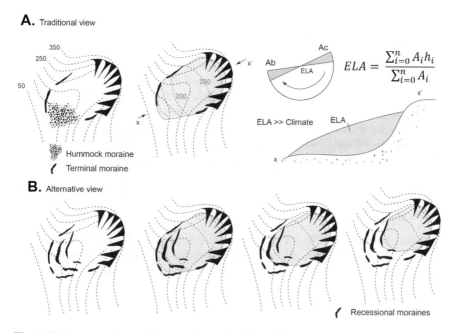

A. Traditional view

B. Alternative view

Hummock moraine

Terminal moraine

Recessional moraines

$$ELA = \frac{\sum_{i=0}^{n} A_i h_i}{\sum_{i=0}^{n} A_i}$$

ELA >> Climate

Fig. 11.6 Reconstructing small cirque glaciers. (**a**) The traditional view based on Sissons and his students. They map the extent of a glacier using their landform model (terminal moraines + hummocks). Drawing a margin for the glacier comes next and then they contour the surface by linking the intersection of a contour line with the glacier margin. They draw in the glacier contours by eye bearing in mind that most contours are convex in the ablation zone and concave in the accumulation zone. They then determine the Equilibrium Line Altitude (ELA) using the empirical equation shown where A_i = the area of the glacier surface at contour interval I expressed in km2; hi = the altitude of the midpoint of the contour interval I; and n + 1 = the number or contour intervals. The height of the Equilibrium Line is linked to climate, higher when warmer and drier, lower when colder with more snowfall for example. (**b**) Alternative view in which these glaciers are seen to decay actively, you can work out the ELA through time as the glacier decays

equilibrium line altitude (ELA) was based on area (Fig. 11.6a). By mapping lots of ELA's across the country you can get a handle on what the climate was like during the Younger Dryas. They even went as far as building a synoptic weather map of the depressions causing the snow fall. Using this simple model large areas of the Highlands were mapped.

All was good until a Mancunian lad turned up in Edinburgh under the supervision of G.S. Boulton in the late 1980s. Instead of seeing chaos in the hummocks the saw order, order which could be interpreted courtesy of modern analogues in Norway and Iceland. They put forward the idea that instead of stagnating the Younger Dryas glaciers decayed actively, retreating one moraine at a time (Fig. 11.6b). They were able to map this pattern of decay over large areas of the Scottish Highlands. The reception to this new model was hostile, academics do not like change to established paradigms, and can be unkind, harsh, and brutal to those suggesting alternative ideas. And this idea was quite radical, suggesting that despite the rapid rise in global climate these mountain glaciers were able to survive for much longer and decay actively. Why? Because they modified their own climate. Glaciers cool the air above them which produced high pressure and one of the results is katabatic winds, that is winds that flow from the cold interior towards the glacier margin keeping it cool. So close to the margin conditions may remain cold but a few miles away they may be much warmer so the beetles from a bog a few kilometres away may show warm temperatures, but it may still have been frigid at the ice margin. The significance is that it shows how today glaciers may respond in quite complex ways to climate change. It took perhaps 10 years for the paradigm to change, many of which were very painful for the young lad from Manchester, but the idea is now accepted, and people map active decay. That young lad was me.

- Task 11.3, to find out the current state of this debate there is a good review paper by Bickerdike et al. (2018) which sets out the glacial landsystems for the Younger Dryas in Britain and also the implications for deglaciation.

Before we leave the British Isles, we just need to mention those areas outside a glacial limit, one of which off course is Dorset. These areas have been subject to intense periglacial action on multiple occasions during the Quaternary. One of the features of modern periglacial terrane in the high-Arctic is the presence of large lens of segregated ice. In permanently frozen ground water can migrate towards the freezing front where it becomes frozen building up into large lens of ice within the body of the soil or rock. When this melts out it creates thermokarst, a complex terrain of enclosed subsidence basins and adjacent ridges. Flat lying areas may also be subject to thermal cracking and if water is available seasonally to fill these cracks, they can become vertical ice wedges (Fig. 11.7). In the winter they contract again and this in turn becomes water and then ice filled. When climate warms and the ice melt the wedge once filled by ice is infilled by sediment from above. Fossil examples of ice wedges are to be found throughout the home counties of the UK in the wall of gravel pits. Given that you need quite specific climate conditions to cause the thermal cracking their presence has been used to give an approximate idea of winter conditions.

Patterned ground, stone circles and stripes are a common feature of modern periglacial environments (Fig. 11.7). They involve both a sorting and a pattern process and are not as well understood as some land forming processes. The sorting process involves the upward movement of stones in seasonally frozen ground. Think of an elongated stone sitting vertically in a soil. As the ground freezes in the winter from the surface down the freezing front will encounter the top of the stone first. The stone will conduct the cold better than the surrounding soil which has air voids in it. As a result, a small lens of ice may build up beneath the elongated stone. The soil water is drawn to the cold stone and freezes forming an ice lens that raises the stone up. When the ground thaws loose soil fill the void left by the ice lens former ice before the stone can settle back, in that way it is lifted slightly. Over time the stone works its way to the surface and then be moved by local slopes. For example, it the surface is slightly domed the stones will roll to the margins.

Relict periglacial features are to be found in upland areas throughout England and Wales. Exmoor has some particular fine examples, and the mountain tops of

Fig. 11.7 Periglacial features. (**a**) Ice wedges, as the ground contracts it fractures into a polygonal pattern. (**b**) Ice wedge note the vertical banding in the ice. (**c**) Relict frost wedge in a gravel pit in Oxfordshire. (**d**) Patterns ground in Svalbard. (Part A is curtesy of Emma Pike Public Domain, https://commons.wikimedia.org/w/index.php?curid=2841420. The other images are from the authors collection)

Wales have periglacial features dating from the Younger Dryas and perhaps the Little Ice Age. Periglacial features are also still active today on some Scottish mountains. One of the most widespread of all periglacial processes is solifluction which is a type of mass movement facilitated by seasonal ice. Large amounts of sediment would have been mobilised in this way during periglacial periods and is referred to as 'head'. Most coast exposures along the south coast have exposures of head. Mobilisation of large mass movements such as those on the Dorset Coast would also have been facilitated by solifluction. Linked to this is the process of cambering which is widespread across southern Britain. This occurs where a more competent jointed rock overlies a clay. A common situation around the Weald or in the Mendips.

Under periglacial conditions the overlying competent rock moves towards the valley floor opening up cracks and fissures behind the moving blocks. These fissures are referred to as 'gulls' and quite extensive caves systems can be opened up in this way. They can be a particular hazard for building.

Perhaps the most romantic of periglacial features is the pingo, which is an Inuit word for ice hill (Fig. 11.8). This is exactly what it is. If we get a shallow lake or saturated zone on a flood plain that then begins to freeze the ice encroaches from the outside creating a large ice blister which can raise and deform the surface. This is an example of a closed pingo, there is a finite amount of ice in the system. Alternatively, where a spring outcrops on a valley floor there is a continuous supply of water which may freeze close to the surface to create an ice blister in the same way. This is an open system since it can grow as long as the spring continues to feed water. When climate gets warmer the ice lens melts and the surface collapses to create a circular rampart. Remains of pingos have been described throughout parts of Norfolk and Suffolk and even in mid-wales. Perhaps the most intriguing examples are beneath London where deep enclosed hollows have been interpreted by some as relict pingos. These hollows caused problems for Victorian engineers digging the London Underground. So, in summary there is a range of features that can be found outside the glacial limits in the UK and the features of a periglacial landsystem are summaries in Fig. 11.9.

Amazonia

Given that many of you on the course are ecologists and others may want to look beyond the shores of the UK I thought we would briefly look at the Amazon Rainforest from a geological perspective. The Amazon Rainforest is held up as one of the most important ecological features of our planet and there is no doubt that it has a role in ventilating the tropics as discussed in Chap. 5. It also has a remarkable biodiversity. Traditionally this has been explained by the idea of refugia. A refugia is where plants and animals survived the rigors of Cenozoic climate change

Fig. 11.8 Pingos. (**a**) Active pingo in the Northwest Territories of Canada. (**b**) Partially melted pingo in Reindalen, Svalbard

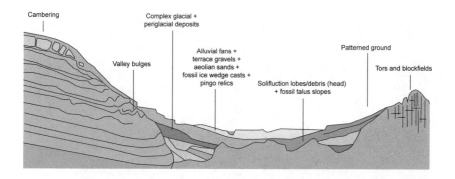

Fig. 11.9 Features of the periglacial landsystem. (Eyles, N., Paul, M.A. Landforms and sediments resulting from former periglacial climates. In Eyles, N. (ed) Glacial Geology, Pergamon Press, Figure 5.1)

emerging periodically during warmer climates. This is now known to be wrong, and we need another explanation for the shear diversity of faunal and flora in this region, a diversity that surpasses other tropical forest on Earth. So why?

- Task 11.4, check out the paper by Hoorn et al. (2010) on which this section is based.

The answer is complex but lies ultimately with sediment supply and the soil (Hoorn et al., 2010). Most tropical forest soils are deeply weathered, all that warm water favours weathering, and they survive primarily because of rapid recycling of falling biomass. The exception is the Amazon Basin which has some of the most diverse soils of any tropical region and the reason is due to the Andean Orogeny. Figure 11.10 shows a series of palaeogeographic maps of the Amazon Basin through the Cenozoic and the slow transition from a landscape dominated by a weathered craton to one dominated by sediment supply from the Andes. The drainage basin developed progressively initially with water flowing north through Columbia before a series of giant wetlands developed and ultimately the drainage adopted its current route out to the Atlantic. Throughout this time the Andes where being pushed up leading to a continual input of rock debris much of which was trapped in foreland basins. Foreland basins are basins in front of a growing mountain range, in this the case on the eastern side of the mountain range. Lots of different rock types were involved creating a diverse soil system. It is this nutrient rich soil that has driven the biodiversity we see today. Reduced to its crudes elements plate tectonics is responsible for the Amazon Rainforest and its huge biodiversity (Fig. 11.11).

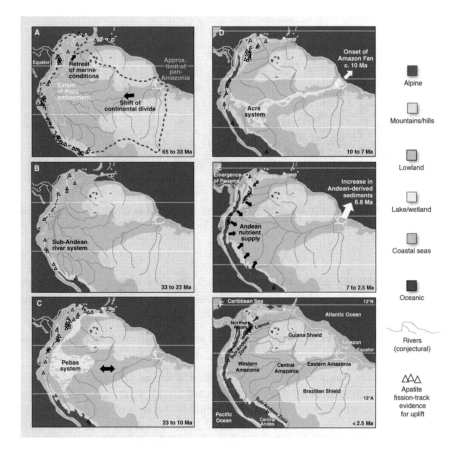

Fig. 11.10 Series of palaeogeographic mpas showing the evolution of the Amazon Basin during the Cenozoic. (Reproduced from: Hoorn, C., et al., 2010. Amazonia through time: Andean uplift, climate change, landscape evolution, and biodiversity. *Science 330*, 927–931, Figure 1)

Summary

In time I hope to add more case studies to this chapter but for now we will have to call it quits at this point. I hope the primer has given you some foundation in physical geography, there is so much more I could write and perhaps will in the future. When you look at the landscape around you in the future, I hope you will see the clues in the landscape that will tell you about the process and timescales over which the landscape has formed. The ecologists amongst you I hope that you will look below the plants and the wildlife and see the rocks, sediments, and soils below, without this foundation life would not be the same.

It has often been said that those who see and tackle the big questions are those outside the core discipline. Take Alfred Wegener for example, of continental drift fame, he was a meteorologist by trade yet his contribution to geology (perhaps not

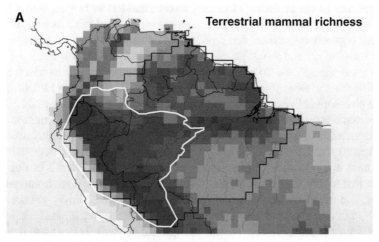

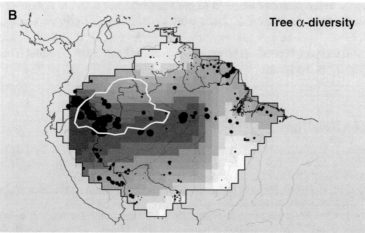

Fig. 11.11 Present Amazonian diversity patterns. (**a**) Terrestrial mammal richness (range: lightest colour, 2–10 species; darkest, 89 to 109 species); white polygon denotes relatively rich soils. (**b**) Tree α-diversity (66). Black dots: local tree α-diversity on 1-ha plots (n = 752); Fisher's α ranges from 3.6 to 300; green shades: loess spatial interpolation of 1-ha values (6–117); white polygon: area of least severe water shortage. (Reproduced from: Hoorn, C., et al., 2010. Amazonia through time: Andean uplift, climate change, landscape evolution, and biodiversity. *Science 330*, 927–931, Figure 3)

in his lifetime) was far greater. If you live firmly within the paradigms and norms of a single discipline it is hard to think heretical thoughts and to question fundamental principles. This is one argument for the power of multidisciplinary research and education.

Natural philosophy is rather a dated term but is one that is powerful in an age where cross-disciplinary science holds many of the answers. It derives from the Latin philosophia naturalis and is the study of nature and the physical universe. It is considered to be the precursor of natural science and has its origins with Aristotle. It has been superseded by the modern concept of 'science' with its multiple often isolating pigeonholes such as ecology, biology, chemistry, physics, geology, geography, archaeology and so on. Yet our world is holistic and earth systems are linked across many disciplines. To understand these systems, one needs to take a holistic multidisciplinary view, just as Alfred Wegener did. So, I like the idea of being a natural philosopher because it stresses the value of inter-disciplinary study essential to understanding a holistic system. My first degree is in Physical Geography and I come from a line of geographers, but in truth I have worked across many disciplines in the last 30 years. Students like to identify with a subject: we are Geographers why do we need to do chemistry, or I am an ecologist why do I need all this geology? The truth is that to understand the natural workings of our physical environment, past, present, and future we need a broad grounding in multidisciplinary science and to have the tools to communicate with other specialists. So rather than letting your chosen degree pigeonhole you into one discipline think broadly as a natural philosopher would and remember we need to see the world not with the arrogance of our own needs but as middle-aged planet with a rich history of climate change, plate tectonics, sea level change, extinctions, and evolutions. And as any middle-aged planet will tell you, it needs to be cared for, nurtured, and valued so that it can have a rich history in the future.

Further Reading

Bickerdike, H. L., et al. (2018). Glacial landsystems, retreat dynamics and controls on Loch Lomond Stadial (Younger Dryas) glaciation in Britain. *Boreas, 47*, 202–224.

Clark, C. D., et al. (2004). Map and GIS database of glacial landforms and features related to the last British Ice Sheet. *Boreas, 33*, 359–375.

Evans, D. J., Clark, C. D., & Mitchell, W. A. (2005). The last British Ice Sheet: A review of the evidence utilised in the compilation of the Glacial Map of Britain. *Earth-Science Reviews, 70*, 253–312.

Stokes, C. R., et al. (2016). Ice stream activity scaled to ice sheet volume during Laurentide Ice Sheet deglaciation. *Nature, 530*, 322–326.

Hoorn, C., et al. (2010). Amazonia through time: Andean uplift, climate change, landscape evolution, and biodiversity. *Science, 330*, 927–931.

Printed in the United States
by Baker & Taylor Publisher Services